MÄRKLIN

1895 - 1914

First published 1983 in the United Kingdom by
Denys Ingram Publishers, PO Box 287, London N1 1EU
ISBN 0 907724 02 7
Limited edition of 200 copies:
ISBN 0 907724 03 5

First published 1983 in France by
Editions d'art Monelle Hayot, Paris
ISBN 2 903824 04 5

First published 1983 in Austria, Germany and
Switzerland by Orell Füssli Verlag, Zürich
ISBN 3 280 01427 1

Consultants
Gilles Hervé / David Pressland / W. E. Finlayson
Editor Charlotte Parry-Crooke
Research Charlotte Parry-Crooke / Justin Knowles
Production Consultant Vanessa Charles
Photography Graham Strong
Design Martin Groot
Filmsetter KNP Group
Printer de Lange/van Leer bv, The Netherlands

The publishers gratefully acknowledge the kind help of all those who have contributed to the compilation of this book. They would particularly like to thank the following:
Gilles Desmons, French translation
Michael Roesen, German translation
Pierre Boogaerts / Charlotte Breese
Allen Levy, London Toy & Model Museum
Denis Ozanne / Christopher Pick / Piko
Claude Wanecq / Brian Webb / Felicity White.

MÄRKLIN

1895 - 1914

Introduction by Charlotte Parry-Crooke
Principal Consultant, Gilles Hervé

Denys Ingram Publishers, London

Foreword

The outstanding German toymaking company of Märklin is the subject of this first volume of a major new series devoted to the great toys and toymakers of the late 19th and early 20th centuries.

Most Märklin toys today belong to private collectors; they are rarely, if ever, seen – even by other toy enthusiasts. Until now, their magnificence has been hidden behind closed doors; the only readily accessible references to these important objects have been the stylized black-and-white line drawings contained in reprints of the factory catalogues.

Märklin 1895-1914 is the first publication to show in classic colour photographs a wide selection of fine Märklin toys, almost without exception in original condition, and all produced during the peak period of the firm's history. Most of the pieces illustrated come from two major distinguished private collections.

This book does not, however, attempt to be definitive or comprehensive in the range of toys it shows or in the information it provides; it aims instead to reflect the manifold qualities of the fine products issued by Märklin during 'The Golden Age of Toys' and to highlight the aesthetic and technical ideals and standards of their creators.

And it is hoped that every reader, specialist and non-specialist alike, will find interest and pleasure in the vivid social record of early 20th-century inventions and developments (especially in the transport area) captured for posterity in three-dimensional form by these toys.

I should like to thank everyone who has helped in the compilation and production of this book; my special appreciation and thanks to Gilles Hervé, Charlotte Parry-Crooke, David Pressland and W. E. Finlayson for their informed, sensitive and generous co-operation.

Justin Knowles
Denys Ingram Publishers, London

	Contents	Sommaire	Inhalt

THE EARLY HISTORY OF GEBRÜDER MÄRKLIN

The great tin toymaking firm of Märklin is the subject of this book. The Märklin company, despite significant competition from rivals in the Nuremberg area, was and continues to be considered the 'king' of the German toymakers. From 1888, when the firm became known as Gebrüder Märklin, it expanded steadily, becoming by the years immediately following the turn of the century the major supplier of mechanical toys to the international market.

Märklin's wares included superb ranges of locomotives, stations and accessories, automobiles, boats, aeroplanes and novelties – each individual item impressively assembled and painted by hand to the highest standards. The firm's business acumen, its readiness to embrace new merchandizing methods and its ability to cater for specific markets, coupled with its excellent products, brought outstanding international acclaim.

The reputation the company made for itself in the early years of this century remains unchallenged; the toys produced in Märklin's peak period – 1895-1914 – are the summit of the aspirations of today's collectors.

Tin Toymaking Centres in Southern Germany

The climate that prevailed in Europe at the turn of the century was one of optimism. A flood of new inventions had caused remarkable leaps forward in methods of manufacture and communications; and the buying power of the public had increased enormously. These developments furthered the swift expansion of most industries. The tin toymaking business was no exception: in its history the years between 1880 and 1914 are referred to as 'The Golden Age of Toys', an appropriate description and one that is especially relevant to the tin toys produced in these years by the numerous manufacturers in the Nuremberg and Württemberg areas of Germany, both traditional centres of toymaking.

The industry in and around the city of Nuremberg was geographically compact and concentrated, in spite of the large number of firms based there; the businesses of Württemberg, including Märklin, were dispersed throughout the state. The reputation of the Nurembergers spread as a whole; in contrast, the Württemberg companies were known individually, independent of any collective reputation. This was due more to the individual spirit of the Swabian people of Württemberg than to the lack of a main production centre. In terms of toymaking, the Swabians went their own way, developing their own ranges, production methods and marketing systems, remaining independent yet aware of the progress of their Bavarian rivals in Nuremberg.

In both areas tin toy businesses had almost invariably started as small family concerns. By the early 1900s some makers had achieved world-wide recognition, others moderate success; several had already by this time been absorbed into more rapidly expanding firms. The demise of most companies came in the 1920s and 1930s as a result of the turmoil of the Great War and the ensuing depression. Of the many great Nuremberg firms established in the late 19th century only two or three survive in some form today; Hess, Plank and Issmayer all closed down in the 1930s, as did the 'giant' of Nuremberg (and Märklin's main rival), Gebrüder Bing.

Of the Württemberg companies, only the Märklin business has sustained continuous production from its formation in 1859 to the present time. Extraordinary tenacity and determination to succeed and an uncompromising dedication to quality have ensured its existence. From very modest beginnings, through years of financial uncertainty and precariousness into those of growth and stability, it has built and held its international position, standing alone in its field today as the symbol of over a century of great German toymaking.

The Märklin business was based on the loyalty and commitment of the Märklin family: from the first generation of toymakers to the third (the founder's grandson was still active in the company on his death in 1961), an unflinching will to survive is the theme that has recurred time and again in the history of both family and firm. Much of our knowledge of the early years of the Märklin story comes from the family chronicle compiled by Wilhelm Märklin (1900-63); this is based on his historical research and on information supplied by Eugen Märklin (1861-1947), son of the founder of the business and initiator of the measures which led his father's creation to eventual fruition.

From the 14th century the Württemberg area had been the home of the Märklin family; the genealogy is well recorded since many descendants either went into the Church or became pharmacists. In 1466 the family received its coat of arms: a gold lion holding an hourglass displayed on a blue field. At this period Nikolaus Märklin (born *c.* 1430) of Donauworth is recorded as a Burgher of Marburg; a direct line continues from him down to the founder of the toymaking firm: Theodor Märklin.

The Formation of the Märklin Toy Business

Theodor Friedrich Wilhelm Märklin was born in 1817 and died in 1866. From the age of three he was brought up in an orphanage. On leaving he became a plumber's apprentice and later a Master Plumber. By 1840 he had set up in Göppingen. Four years later he married a local girl, Rosine Geiger. They had two daughters: Rösle (who eventually emigrated to the USA) and Sophie (who later played a small part in the business; her husband, Gottfried Britsch, joined Eugen Märklin when he saved the firm from collapse in 1888).

Municipal citizenship was granted to Theodor in 1856; he also received his Master's Certificate as a sheet metal worker that year, so he had obviously changed his profession. In 1857 Rosine died, leaving Theodor to run the business and care for the children on his own. Two years later, in 1859, he remarried. His second wife was Caroline Hettich from Ludwigsburg (1826-93), a relation on her mother's side of the composer Liszt. A strong and courageous woman, Caroline was to be largely responsible for the survival of the business in its early years.

Shortly after their marriage the couple embarked on the production of metal toys for dolls' house kitchens; this was presumably a natural progression from Theodor's sheet metal work. Thus was founded the original Märklin toymaking business. The company was known as W. Märklin; it traded under this name from 1859 until 1888. From the outset Theodor and Caroline ran the business together: he was responsible for production, she for sales. Caroline was one of the first, if not the first, female sales representatives. Her selling and marketing talents took her throughout Germany and Switzerland obtaining orders for toys. Dedicated to the expansion of the business, her intention was to build it up for her children to inherit.

Between selling trips Caroline found the time to produce three sons: Wilhelm (born 1859), Eugen (1861-1947), who became the key figure in the future history of the company, and Karl (born 1866). With the two daughters from Theodor's first marriage there were now five children to support. But the selling trips were fruitful and after several years a move to provide increased work space became necessary. Number 56 Grabenstrasse was acquired: from here both business and family life were conducted. A proper workroom and storeroom were available and an apprentice was employed.

In 1866 unforseen circumstances halted the progress of the business, however. Theodor met with an accident which caused his death at the age of 49. His apprentice had left a trap-door to the cellar open one evening; Theodor fell into the cellar, broke several ribs and subsequently contracted pneumonia from which he died. His youngest son, Karl, was six months old at the time. Caroline was left with the responsibility of the upbringing of the five children and the running of the business.

Despite her laudable determination, trade did not flourish. In 1868 she herself remarried, partly in an effort to bring fresh energy into the deteriorating firm. But this relationship was a dismal failure on all levels.

The second husband provided none of the much needed assistance; things were so bad that the Märklin sons were sent to grow up away from home. They only returned on their step-father's death in 1886.

This latter event left Caroline once again in charge of the business; despite the seemingly never-ending struggles, she refused to admit defeat. Eugen was given bereavement leave from his stable and remunerative job and during this time he managed to ease some of the accumulated problems. He had done some travelling for his mother between 1881 and 1884, so he was well aware of the precarious situation. Though workers were hired to finish off uncompleted orders, the future of the company was in doubt. Caroline's main motivation had always been her desire to safeguard the business for her offspring, but they now had their own lives which they were hesitant to relinquish. Eugen, particularly, with a good job and the possibility of marriage to Bertha Christianus, was torn between the stability of his current situation and rallying to the aid of his mother.

The Märklin family characteristics of loyalty and commitment prevailed. On March 1st 1888 Eugen and Karl set up a private firm in which their parents' business was incorporated. The new company was known as Gebrüder Märklin (Märklin Brothers). Gottfried Britsch, their half brother-in-law, moved from Stuttgart to Göppingen to provide production expertise. Britsch left after a year, however; arrangements had not worked out as planned.

The New Company

After the unsatisfactory year with Britsch, Eugen became the new firm's leading light. From 1888 to 1892 the business traded as Gebrüder Märklin. In the family chronicle Eugen describes the struggles of the first years, the weighty responsibility and his grave worries. The first financial year ended with a turnover of 10,000 Marks but a loss of 3000. Voluntary liquidation was only averted by increasing the mortgage on the premises.

Caroline Märklin was still involved and was shortly joined by another woman of similar qualities. Bertha Christianus married Eugen after she had completed a year's business apprenticeship; thereafter she played an important role in the running of the firm. This enabled Eugen to concentrate on the sales trips and to accept agencies for toys and household goods. The latter were produced to order to support the slack periods in the toy trade; toys were a seasonal line, the emphasis placed on Christmas. Times were still hard and money for the wages was often almost impossible to raise.

According to the family chronicle Eugen took over marketing the fine tin toys of Ludwig Lutz of Ellwangen in the summer of 1891. He was so encouraged by the popularity of the Lutz toys in northern Germany that he decided to try to buy Lutz out. Eugen states in the chronicle that no down-payment was in fact involved; apparently Lutz even provided a loan of 5000 Marks to enable production to continue. The loan was given on the condition that Eugen looked for a partner to inject capital into the business (and presumably thereby ensure the survival of the Lutz range).

This take-over brought about the next phase in the history of the Märklin company. In accordance with Lutz's stipulation, and a sensible measure under the circumstances, a new partner was found. Emil Friz (1858-1922) from Plochingen joined the firm in 1892. Friz obviously provided the expected capital; accordingly, the new company catalogue announced 'Owners: E. Märklin & E. Friz'. The company name was adapted to Gebrüder Märklin & Co; trading continued under this name until 1907.

Friz showed total dedication to his new commitment; Eugen tells that Friz was determined to make the company the biggest and best toy manufacturer in the world. Some indication of the extent to which he realized his dream can be seen in the illustrations shown in this book. Friz's role was to develop the business; under his auspices steady expansion occurred, despite continuing financial pressures and the need for 14-15 hour days, worked in the effort to fulfil Christmas orders.

Caroline Märklin died in 1893. The death of this matriarchal woman had a demoralizing effect on the firm, but success was in sight and the growth of the business continued. The company transferred in 1895 to larger premises at 21 Marktstrasse, Göppingen. A further move took place in 1900 to a newly constructed building of 6000 square metres in the Stuttgarter Strasse. By 1900 prizes were being won abroad and export ranges were in production. In 1904 the long established firm of Rock und Graner Nachfolger from Biberach was absorbed into the quickly expanding Märklin empire.

Over Christmas 1906 Eugen contracted a severe spinal inflammation, the effects of which lasted several years. It seems possible that this ailment curtailed his activities somewhat, for in 1907 the next period of the company's history commenced with the arrival of another crucial partner. Richard Safft (1879-1945) joined on May 1st 1907. His role was to expand the export ranges and the foreign trade; his well-documented gift for languages was a great asset.

Eugen Märklin remarks in his notes that from this point the firm enjoyed more freedom and flexibility to act because of a considerable capital increase. This was presumably provided by Safft, for the company name underwent a further alteration in 1907, the year Safft arrived. The catalogue issued this year bears both the former name and the new: Gebrüder Märklin & Cie, indicating the incorporation of another partner. This version of the name was used until the 1920s. Safft's contribution was fruitful: the 1907 catalogue supplement shows a significant widening of both range and production methods.

In 1910 Märklin won the Grand Prix (jointly with Steiff) for the display at the Brussels World Exhibition. By 1911 a further extension of the premises became crucial. A six storey 'high-rise' building, 110 metres long, was constructed next to the single storey factory. By 1914 the number of employees had risen to 600 – a far cry from the one apprentice employed by Theodor.

'The Golden Age of Toys' came to an end with the outbreak of the First World War in 1914; this event called to a halt the seemingly endless stream of creativity from the toy factories of Germany. And it determined the future existence of many toymakers. Though officially outside the scope of this book, it is important to record briefly the changing face of the Märklin company after 1914.

During the war Märklin, like other toymakers, suspended toy production in order to concentrate on the manufacture of armaments. The business survived the war and the ensuing period of inflation under the continued management of Märklin, Friz and Safft. Toy production was resumed after the war; the 1919 catalogue was extensive, though it consisted mainly of pre-war models, which had been stock-piled.

In 1922 the status of the company was again altered. Until this date the firm had been a private company: now it became a limited liability company. The previous tax burden was the reason for this change according to Eugen. The alteration was indicated by the addition of GmbH to the company name. During the negotiations Emil Friz died. Despite this loss to the firm, the next generation had been thoroughly trained to take over when the need arose.

After his death, Friz's son-in-law became the third managing director. Eugen himself, the driving force of the firm, retired in 1935, though remaining a co-proprietor until his death in 1947. His son, Fritz, had joined the firm in 1923 after studying mass production in the USA; he was groomed to succeed his father and remained part of the management until his own death in 1961. Richard Safft died in 1945; his son Herbert was made a member of the board in 1949.

Thus was ensured the future of the business built up by the unswerving dedication of Märklin, Friz and Safft. One worker was employed when the firm was founded in 1859; by 1914 there were 600, by 1928 over 900, and by 1959 more than 2000 – some indication of the partners' achievement. Though Fritz Märklin's death in 1961 left the company without a member of the family at the helm for the first time in 102 years, Claudius Märklin joined the company in 1975.

The Take-overs

As has been noted, between the time of the formation of the company and the outbreak of the First World War Märklin absorbed two other businesses which specialized in the manufacture of tin toys: Ludwig Lutz of Ellwangen and Rock und Graner Nachfolger (R&GN) of Biberach. The toys made by both these well established firms during the latter part of the 19th century were imaginatively designed though fragile because of the thin tinplate used in their construction. Märklin's achievement was to make constructive but selective use of certain elements of the Lutz and R&GN ranges, and to improve the quality and manufacture of their products, while at the same time withstanding the organizational and financial pressures which had caused the demise of Lutz and R&GN.

The date of the Lutz take-over was 1891; it was a crucial contributing event in Märklin's subsequent expansion and eventual success. It made available to Märklin the Lutz ranges and developments in coupling, gauge and track systems – and directly caused Emil Friz's involvement in the Märklin company. The Lutz firm had an excellent reputation and Märklin was able to capitalize on this too.

Ludwig Lutz founded his business in Ellwangen an der Jagst in *c.* 1846. Initially he specialized in fine decorative tin work, turning later to concentrate exclusively on the production of tin toys, especially dolls' house accessories and transport toys such as horse-drawn carriages and wagons. In 1883 August Lutz, Ludwig's son, took over the business; with 25 employees he expanded the range, introducing clockwork and steam driven toys. Bing handled the marketing of the Lutz ranges which were popular in northern Germany and abroad.

In the summer of 1891 Märklin took on the Lutz marketing and later that year, in the autumn, absorbed the firm. These dates are given in Eugen Märklin's own notes concerning the take-over; other sources suggest that Märklin had marketed the Lutz toys prior to 1891 (even from the 1870s) and give alternative dating on the introduction of new Lutz ranges and the take-over itself. We can, however, only assume that Eugen Märklin, who himself conducted the take-over negotiations, gives the correct chronology.

Once the terms of the deal were agreed (a loan from Lutz on the condition that Märklin found a partner to inject capital into the firm), production at Ellwangen ceased. The Lutz machinery and tools, together with the stock, were re-installed in the Märklin premises; the Lutz employees were given the opportunity of moving to Göppingen and continuing work there.

The transfer of the Lutz equipment to Göppingen enabled Märklin to take advantage of the developments on which Lutz was working at the time of the take-over. Lutz-designed products were immediately incorporated unchanged into the Märklin programme and more importantly Märklin was able to bring to completion, and market under its own name, current Lutz prototypes on which work was suspended at Ellwangen when Lutz closed down.

This chain of events has given both collectors and historians much cause for confusion over the provenance of Lutz/Märklin toys of this period and their specific dates. No known examples of marked Lutz toys exist, providing a further complication. It is now thought that some toys sold under the Märklin banner were in fact made by Lutz prior to the take-over; others were undoubtedly produced afterwards, but on the original Lutz machinery and possibly by Lutz workers at Märklin. Even the first train on a sectional track with which Märklin caused such a sensation at the 1891 Leipzig Fair is now thought to have been a Lutz model; the hand-coloured Lutz sample books (in Märklin's possession) give an indication of the origin of Märklin's first trains on tracks. According to one contemporary source, Lutz had produced clockwork and steam trains running on rails by 1889.

Märklin also made use of Lutz's coupling systems and the firm's work on the three larger gauges; Märklin was the company that introduced the gauge classification system which was universally adopted at the time and which is still in use today. To what extent Lutz was indirectly responsible for some of the innovations attributed to Märklin can only be assessed hypothetically at the present time, but it was from the date of this take-over that Märklin began its ascent towards unparalleled success in the toy and model railway field. We know too that Märklin itself recognized its debt to Lutz in terms of Lutz's contribution in both product and reputation – some indication of the long-term influence of the take-over on the Göppingen firm's development.

The early history of Rock und Graner is rather better documented than that of Lutz, though areas of doubt still exist over the extent of Märklin's involvement in R&GN's liquidation in 1904. Probably the earliest Württemberg firm to manufacture tin toys, Rock und Graner was founded in Biberach an der Riss in 1813. Its fine 'lacquered' sheet metal toys are first referred to in 1837; in this same year over 100 people were employed. In 1851 the firm won the Gold Medal at the Great Exhibition in London; from then until the 1870s it enjoyed considerable success with horse-drawn carriages, fire engines, steam carriages and other lines. After the 1870s a decline set in and the founders' descendants lost interest in the business. Products shown at the Württemberg State Trade Fair of 1881 however apparently included vehicles, steam ships, railways and metal dolls' house furniture. In 1896 Oskar Egelhaaf took over the firm; the manufacture of railways and related accessories became the principal line of work. After Egelhaaf's involvement the firm was known as Rock und Graner Nachfolger (successor) and its trademark was R&GN.

By 1900 Egelhaaf's own interest in the firm had waned; only 36 workers were employed that year. Egelhaaf had other business interests he wished to pursue in Ulm; consequently he had no further use for the toy production tools. R&GN was put into liquidation in 1904 and Märklin took over the remaining equipment.

Since Eugen Märklin does not describe in his notes the agreement he came to with Egelhaaf, we have no factual record of the details of the take-over. It has been suggested that Märklin absorbed R&GN in an effort to eliminate a rival; but R&GN's very reduced circumstances at this date can hardly have constituted serious competition for Märklin. In addition, it is not certain to what extent Märklin was involved in the subsequent use of the remaining R&GN assets; it is thought that Märklin acquired only some of the technical equipment and designs, but not any of the models. Yet the similarity of the two firms' railway ranges is often remarked on. No R&GN railway models appear in the subsequent Märklin catalogues, however, though there are examples of R&GN track layouts.

Märklin undoubtedly made selective use of certain elements of the R&GN business, but the acquisition of the latter firm was not as influential on future developments as the Lutz take-over. Nevertheless, it must be said that the way in which Märklin constructively built on the design concepts of both Lutz and R&GN, in conjunction with its own high standards of production, probably constituted the catalyst that brought about the remarkable toys assembled in the Göppingen workshops up to 1914.

The Pre-war Product Range

In the years following the Lutz take-over Märklin's toy range expanded dramatically. Eugen Märklin remarks in his notes that the company was consistently creating new toys based on all spheres of human life and that these became a source of inspiration to the whole toy industry. This is by no means an understatement: in 1895 the main catalogue offered 550 different lines; that of 1902 displayed in the region of 1650 – an extraordinarily ambitious range even by today's standards. Despite this impressive expansion, constant rationalization and systematic development took place in both manufacturing methods and choice of product. In terms of subject-matter Märklin focused its attention increasingly on toys which reflected the various types of transport of the age: locomotives and all types of railway accessory, automobiles and nautical and aeronautical items.

A study of the comprehensive catalogues issued by Märklin between 1895 and 1914 provides an illuminating record of developments as they took place. In the early days items were constructed as simply as possible; later more complicated production processes were used. New mechanisms were incorporated, adapted and refined, making possible even such manoeuvres as the remote control of a boat by means of a rubber tube system. Major developments of new models were dependent on the availability of finance for the necessary tools; hence the 1907 catalogue supplement, produced after Richard Safft joined the firm, shows a definite widening both of the range offered and of production methods.

As Eugen Märklin pointed out, the inspiration of the Märklin toys was real life. Most toymakers of the period looked to reality for suitable subjects to reincarnate in toy form, but Märklin was more daring than most. The pages of its catalogues are filled not only with immediately acceptable items of guaranteed appeal; in an effort to provide the most extensive range possible the company was prepared to take risks with such improbable items as snow ploughs, central heating railway cars and Red Cross carriages, to name but a few. The attempts to provide the ultimate toys did not end here; a sharp eye was kept on contemporary inventions and enthusiasms and Märklin was often the first to reproduce the latest craze in miniature.

This awareness extended to events abroad, especially countries where sales were particularly high, notably Britain. Export ranges were constantly expanded and enormous efforts made to provide suitable toys for the different national markets. Locomotives and carriages were available in a wide variety of different liveries and boats supplied with appropriate national names and flags. And suggestions from clients were welcomed and often put into production, as in the case of Mr A. W. Gamage, the London distributor for many manufacturers, including Märklin. Individual catalogue supplements for the main foreign markets were issued in the years leading up to the First World War.

The foreign buyers were particularly interested in Märklin's transport lines; this was the range which brought the firm international repute. Other lines continued to be produced, however, in smaller runs but with equal flair. And the dolls' cooking stoves of the very early days remained on offer. Horse-drawn carriages of all descriptions, similar to those made by Lutz, were popular buys in the 1890s and 1900s, as were prams, merry-go-rounds and fortresses. Working fountains and windmills were available from 1895; castles and delightful miniature swimming pools were featured in the 1902 catalogue. All types of fire engine and fire fighting equipment were on offer too; these had great appeal and were superbly made in the 1900s. Perhaps the most beautiful of all Märklin's novelties was the 1909 'Fidelitas' circus road train – a cavalcade of clowns bedecked with flowers and gorgeously costumed (see pages 172-3).

Despite the attractions of these wonders, Märklin chose to concentrate on the development of the transport ranges, so admired by the foreign buyers. It is worth considering briefly therefore a few of the firm's achievements in the different transport areas. Locomotives and their accessories rapidly became the major line of business. Märklin had concentrated on floor trains before 1891. After the Lutz take-over that year the first train to run on a sectional track was presented at the Leipzig Fair. In the same year the Märklin gauge classification system was introduced, examples of gauges 0 and 1 being shown at Leipzig; this system was adopted universally and thereafter Märklin offered most of its models in a choice of several if not all the gauges. In 1892 the first clockwork railway with a track in the form of a figure '8' with points and crossings was presented; this also caused a sensation when it first appeared.

Between 1891 and 1895 improvements were made to the existing locomotives, including the Lutz models; the range was constantly expanded to cover all gauges. Carriages, on the other hand, remained virtually unaltered until 1900. In 1895 structural changes started to occur and more emphasis began to be put on the export ranges, with attention given to individual liveries and characteristics, though it was not until later that models appeared in foreign, as opposed to German, outline. In 1898 Märklin achieved yet another triumph with the issue of the first electric train sets made in Europe. By 1907-9 the introduction of new production processes, such as die-stamping, had facilitated yet further expansion of the range. In 1909 Märklin introduced Coupling 5, known as 'Fix'; between 1891 and 1915 many new types of coupling were brought in, but 'Fix' became the universal coupling and was in constant use until 1960.

Among Märklin's most impressive toys are those of nautical inspiration: yachts, paddle-boats, liners, passenger ships, battleships and so forth. These were produced from the mid-1890s and run by clockwork or steam. Their quality is consistently fine and they are without exception extremely attractive toys, beautifully made and finished to very high standards. In this field, too, Märklin reigned supreme, though competition was provided by Bing until the early 1900s when the latter's nautical toys declined to some extent. New models were brought in by Märklin at intervals but in general the boats changed little until the 1920s. Until this time the proportions of the entire nautical range were charmingly exaggerated; Bing's boats, by contrast, were more exact representations of their real-life counterparts.

Märklin was less successful in the face of competition from Bing and also Carette in the field of automobiles, though the Märklin examples were of the usual high quality. Märklin was, however, the first tin toymaker to introduce a range of automobiles. This was in 1900; Bing first offered cars in 1902 and Plank only in 1904. Surprisingly in this area even Märklin was somewhat slow to react. The first Daimler and Benz cars had appeared in 1886, followed by models from other makers, such as Peugeot and De Dion, in the early 1890s. But by 1904 Märklin offered a wide range of automobiles – though somehow they never quite matched the excellence of the Bing and Carette examples. Production of high quality automobiles was continued after the war by Märklin, while Bing ceased to offer such items in 1914.

In the field of aeronautical toys Märklin was impressively quick in capitalizing on the public's enthusiastic response to another revolutionary mode of transport. Following Wilbur Wright's successful demonstration of his biplane in France in 1908, Märklin issued that same year three Wright Brothers biplanes in different sizes, as well as a selection of airships, including Zeppelins. Only by 1912 did almost all the rival manufacturers offer similar items, except for Müller & Kadeder, whose 1909 catalogue featured both biplanes and Zeppelins. Märklin produced several truly superb aeronautical toys; perhaps the most impressive is the largest version of the Wright Brothers biplane with its elegant swan's neck tiller and powerful clockwork motor (see pages 156-7). But for some reason the aeronautical range was not expanded greatly, and consequently these toys are very rare nowadays.

The Qualities of the Pre-war Toys

In all the fields described in the last section, Märklin established a unique position. The wide ranges and many important innovations certainly contributed to the firm's success. But why, one might wonder, does each individual Märklin toy stand out from its counterparts offered by the company's competitors. The answer is to be found in Märklin's unparalleled attention to quality; from concept to design, and through all stages of production to completion, extraordinarily high standards were set and maintained.

As has been pointed out previously, in terms of subject-matter Märklin showed a great deal more imagination than most other toy manufacturers. The firm's readiness to follow its own creative instincts, and its imaginative and distinctive use of source material, brought about the issue of some of the finest toys ever made. The lengths to which Märklin was prepared to go in the pursuit of original ideas have been called perversely idealistic; but we should be grateful for this relentless search for new inspiration, for these toys provide for us today a magnificent record of social history. In more than a few instances a Märklin toy is the only surviving three-dimensional representation of, for example, a long outdated fire engine or a demolished station.

In design, too, Märklin toys are distinctive and a definite 'house style' can be observed. Toys made by Märklin have a solid quality which makes them immediately distinguishable from the products of rival companies. This solidity of form was due both to the exceptionally heavy tinplate used in their construction and to the extended use for many years of the German outline. Other makers, notably Bing, adopted foreign outlines much earlier; Märklin was mostly content in the early years to modify its basic forms for the export ranges.

Exaggerated proportions are another notable hallmark of Märklin toys issued before 1914. Both Bing and Schoenner produced toys which were much more realistic and faithful to the real-life originals. Despite their accuracy, these toys often lack the flair of the Märklin examples. With their exaggerated emphasis of special features, the Märklin toys are much less clinical in feeling, much more exuberant and evocative of their original inspiration, and consequently all the more charming and attractive.

Märklin encapsulated the *spirit* of the original; no attempt was made to scale down details to toy form with absolutely precise accuracy, for in the early days efforts were not made to produce models, just toys. Märklin's depictions were, however, not mere flights of fantasy, based loosely on ideas taken from real trains, boats or planes. Most were in fact faithful in their general appearance, as a comparison of many Märklin toys with photographs of their original counterparts will prove.

In the railway field, as the model train craze grew, Märklin locomotives became increasingly accurate; the company was aware of the potential importance of this movement, which had started in England in the 1900s and soon spread to the Continent, when it became acceptable and fashionable for adults to play with model railways. After the First World War, far greater emphasis was placed on accuracy in all transport toys and Märklin went on to become the manufacturer of some of the finest, most precisely rendered model railways. For the toy collector, however, exact representation is of less importance than the quality of design and production, so the post-war toys are far less appealing than those made before 1914.

In the execution of all the production processes a consistently high standard was maintained; the superb quality that was achieved, and the care that was obviously lavished on each item at every stage, put the toys made by Märklin before the First World War in a class of their own. The attitude was definitely one of craftsmanship as opposed to corner-cutting mass-production, in spite of the vast numbers of toys which had to be despatched from the Göppingen workshops every year. Unstinting attention was given to the quality of materials, mechanisms and construction methods and to the fine execution of the paintwork.

Märklin strove to produce toys which matched their real-life counterparts in quality. Göppingen toys were more expensive than those of many other makers who concentrated on the mass-production of cheaply lithographed and flimsily made items. But the Märklin products were aimed at the upper end of the market and to be able to demand higher prices the toys had to be of good quality. That they were superbly produced, and that they were likely to be looked after carefully, given the type of child for whom they were intended, has meant that those which have survived the years are relatively intact. Thus we are able today to appreciate fully the almost original quality of production achieved by the firm that one collector recently described as the 'Rolls Royce' of toymakers.

Production and Decoration Techniques

During much of the period covered by this book most Märklin toys were constructed and decorated largely by hand. Almost all the different stages through which every Märklin toy passed in its creation were carried out by hand or with the assistance of hand-operated machines; tools and methods remained little changed for years. New, and in some cases revolutionary manufacturing methods, such as offset lithography, which facilitated cost-effective mass-production, were quickly accepted by Märklin's rivals; though there was a gradual transition to more sophisticated methods, for the most part tried and trusted techniques remained in use at Göppingen.

It is difficult to tell retrospectively whether this situation arose by choice or as a result of financial or other pressures. Märklin may well have made a conscious decision to pursue the supply of the quality hand-made toys which were responsible for the firm's reputation; alternatively Göppingen's geographical location, away from Nuremberg and other more concentrated toymaking centres, could possibly have meant that Märklin was less immediately aware of the potential of the new manufacturing methods. Certainly it was almost prohibitively expensive to acquire and install completely new machinery and make complete switches in production methods. Eugen Märklin's notes give few clues as to why hand production was used so extensively and over such a long period; he does however mention that his rivals frequently presented toys at the Leipzig Fair before he could because he had not managed to get the manufacturing facilities at Göppingen adjusted in time.

Whatever the causes of the extended use of hand-executed work by Märklin, its results are, as has been mentioned, of great appeal to today's collectors and give rise to the general view that the creation of a Märklin toy was a process of craftsmanship as opposed to one of mass-production. By what methods was the quality of Märklin toys achieved, how were such consistently high standards maintained throughout, and more crucially how was the production of hundreds of different lines kept economically viable, let alone profitable? Much that happened in the Märklin workshops took place in all toy factories of this date, but a description of what is known of the production, decoration and stock-keeping methods peculiar to Märklin sheds some light on these questions.

Like most toymaking factories the Märklin premises were made up of different workshops; these included the sheet metal works, the tool room, the moulding and pressing works, the assembly room, the precision workshop, the electrical and steam workshops, the painting room and the packaging room. There were also stock rooms and a showroom. Each department played an important role in the creation of a Märklin toy. Much of the labour was highly skilled and experienced; successful production depended on the craft of the individual rather than on the efficient functioning of a machine.

Märklin toys were not, as might be imagined, conceived on the drawing board. Before the days of scale models, an idea for a tin toy was immediately produced in prototype or maquette form. Some sort of reference was probably provided – a newspaper cutting or a photograph, showing the launch of a new liner, for example, or the appearance of a new motor car on the roads.

The first maquette was constructed in tinplate and probably put together mostly from parts designed for previous models. New components, if needed, were cut out by hand from the tinplate with tinplate shears. After adjustments and modifications to this first maquette, a small series of prototypes was prepared for sales purposes. Given a favourable reaction, the prototype was put into production; from maquette stage the toy made its way through most of the different departments in the factory to emerge finely finished, safely packed and ready for distribution.

All the basic components of Märklin toys were made from tinplate; this was prepared for use in the sheet metal works. Traditionally much of the tinplate used by manufacturers of all tin items came from South Wales; this area supplied not only the British industry but also many manufacturers on the Continent. It is possible that Märklin used Welsh tinplate – most of the Nuremberg firms bought from this source – though the Märklin plate is of a heavier gauge than usual.

In the cutting, shaping and assembly departments the traditional tools of the tinsmith were used for many years; these included hand shears, hammers, rollers and soldering tools. Most of the tin cutters were men, though a few women also worked in this department. The basic components of each toy (which often numbered between ten and fifteen separate pieces) were cut by hand or stamped out on hand-fed machines which took only a single sheet of tinplate at a time. Vast numbers of each component were cut or stamped out at each session and then stock-piled for later use, but windows and doors were always cut out individually by hand.

Shapes were usually hammered out by hand over wooden forms; this process was made less arduous by the introduction of pressbenches on which chimneys, domes and other unusually shaped items could be rounded. Embossing was avoided until the mid-1900s; until then three-dimensional effects were achieved through perspective paintwork. Märklin began to use embossing tools in about 1906; by 1908 they were fully operative. The transition to really efficient cutting and shaping methods only came in about 1907 when die-stamping began to be used at Märklin. Evidence of its marked results can be seen in the catalogues issued after this date. Die-stamping revolutionized the production of the basic components of the toys, though a large financial investment was needed initially, since the new tools were extremely expensive. (Possibly Richard Safft's arrival at Märklin in 1907 and the capital he is thought to have injected into the company had something to do with this development.)

Eventually machines run by steam and electricity were introduced into these departments and tools began to be refined more frequently. Many processes, however, were still carried out by hand long after the end of the First World War. Ships' hulls, for instance, were constructed by hand until the 1930s; four sections were drawn over wooden frames, laboriously hammered into shape and soldered together.

The motors for the toys, which were run by steam, electricity or clockwork, were made in house. Some were extremely complex, particularly the steam examples, which involved the use of boilers, pressure chambers, valves and other components; those fitted in the steam-run fire engines and some of the larger boats were especially sophisticated. All motors were extensively tested both before they were inserted in the housings and after the decorative processes had taken place; the latter test was to ensure that no paint had clogged up the mechanism. In the early years the motors were mostly soldered to the housings by hand before the decorative work took place; soldering afterwards would have burnt off the paint. Later on mechanisms were screwed into the housings; this meant that decoration could be completed beforehand.

For some reason in a few cases the old working order continued to be followed: on clockwork locomotives, for example, the motors were screwed into the housings but inserted before decoration, even though there was no special need for this. Generally speaking, however, after soldering had been superseded by screws, most toys were assembled and decorated before the motors were incorporated at the last stage.

The basic components of the toys (body, wheels and so forth) were assembled by a variety of methods. Soldering was common to begin with; subsequently tab and slot systems were used, especially when Märklin finally took to using lithography for the decoration of carriages and cheaper locomotives. If a toy was lithographed, the printing had to be done before the cutting, shaping, embossing and decoration; consequently the printed tin had to be protected as much as possible – embossing tools were covered with felt to safeguard the printed surfaces and since the tinplate could not then be soldered, tabs and slots provided the solution to assembly.

In the decoration rooms hand-painting, spray painting, rubber stamps, stencils and transfers were all used and eventually lithographic printing. Hand-painting remained the most favoured method, however, and the one that was consistently used for decades – it was certainly the technique responsible for the superb quality of Märklin toys. Märklin was one of the few firms to eschew the advantages of lithography in favour of the continued use of hand decoration. Since this was the firm's preferred method of working it is interesting to look briefly at the many separate stages this work involved.

Each part of the painting process was carried out by different teams; one team applied the base coats, another the ground colours, the next the borders and so on. The more complicated the job, the more skilled the labour; most of the very skilled line painters were women. The undecorated tinplate toys were prepared for painting with a de-greasing of hot water, petrol or steam. A base coat or primer was generously applied by hand, or, increasingly after 1900, with a spray gun. The primer covered blemishes in the tinplate and formed a base for the main body colour. A white primer was usually chosen to go under a white top coat, rust under a brown and green under a black. The primed toys were then stoved; this aided the drying process and took place after every stage of painting. The toys were kept in the stoving room, heated to about 40 degrees Centigrade, for varying lengths of time depending on the painting stage just completed. The temperature was high enough to ensure complete drying but not to the point that the paint cracked.

Primed toys were often stored for long periods in the stock room; once primed, little harm could come to them there. When they were required, they were, if necessary, sanded down, re-primed and re-stoved. The main body colour was then applied, again either by hand or with a spray. Another stoving then took place.

Decorative features came next; these included borders, liveries, inscriptions, windows, doors and so forth. Almost all this work was for a long period invariably executed by hand and on the more expensive models until the 1920s. Other methods, such as rubber stamps, stencils, and occasionally transfers, which rationalized the application of the decorative features, were also used.

The most skilled and competent painters were those who were responsible for the particularly detailed and decorative items such as the carousel shown on pages 170-1. As we have seen, their work was especially important in the days before embossing when three-dimensional effects had to be created through clever paintwork. Some of these artists are shown, correctly attired in painters' smocks and carrying palettes and brushes, in a photographic collage of Märklin's Jubilee.

This kind of creative work required a certain artistic flair, but the tasks undertaken by the painters of the more standard decorative features were just as meticulously executed. Each painter mixed his own paints; supplies were controlled by a buyer but each individual prepared his colours to his own requirements. Natural pigments were bound with linseed oil or other resinous mediums and consistency was important – the paint had to be thick and greasy. Borders and outlines were painted freehand; to steady the wrist an armrest, which was the length of the toy being decorated, was sometimes used. The painter rested his arm on the stand and was thus able to move his brush steadily along the length of the toy. The individual nature of this work has meant that it is possible to identify different styles of work. Although names are unfortunately not known, one can sometimes recognize toys which must have been decorated by the same person, given the texture of the paint and the quality of the brushwork.

Inscriptions and the lettering of station names, railway companies, boats and the like were also hand-painted, stencilled on or applied by rubber stamp. The rubber stamp method became especially popular and was also used for details such as portholes on boats, though it only gained favour at Märklin some years after Bing had begun to use it in about 1905. Doors and windows on the larger boats were always hand-painted, even into the 1920s and 1930s. Decorations on the bows or sterns were occasionally applied by transfer, though metal casings were also used.

A good varnish made of natural ingredients provided final protection for the paintwork. Its composition depended on the type of toy it was to be applied to; for instance, an oil-based varnish was the most suitable for steam models since it helped to prevent any possible burning caused by the workings of the steam boiler. The stoving which took place after the varnishing stage lasted longer than the other stovings.

Offset lithography of course meant that all the decorative additions already described needed to be transferred only once on to the litho stone or plate before being machine printed straight on to the tin. Bing, Carette and Gunthermann all adopted lithography as soon as possible and a comparison of the prices of their toys with those made by Märklin shows how comparatively economical this new method was. However Märklin presumably felt that the same standards of quality could not be achieved without hand-painting; litho printing was obviously the cheap alternative to hand-painting, but though a lithographed toy could probably have been produced for roughly 5 per cent of the cost of a hand-painted one, Märklin elected to pursue quality for the most part. Litho printing does appear to have been used at Göppingen in some form from about 1904 but only for the cheaper end of the range. Descriptions in the 1910 and 1912 catalogues give evidence of hand-painted locomotives being supplied with lithographed carriages and by 1913 some of the cheaper locomotives were themselves printed, but none of the steam models.

Packaging of Märklin toys was kept very simple; most items were packed in tough buff-coloured cardboard boxes; heavier items and large sets were supplied in wooden boxes. Ordinarily there were no labels on the boxes, only a reference number, the Märklin name or trademark. Some toys were, however, packed in boxes decorated with lithographed colour labels (see pages 174 and 175) or with black and white line engravings of the contents.

As we know, even in 1902, Märklin was offering in the region of 1650 different lines; multiply this figure by the number of toys likely to have been made up for the run of each line and one might well wonder how production was kept economical, let alone profitable. Two main factors enabled cost-effective production to be maintained: the organization of the company's work force and the impressively efficient systems of stock-keeping.

Jobs were always in demand so there was never a shortage of staff despite the low wages. Productivity levels were maintained and even increased by paying the workers piece rates, based on the quality and quantity of their output – an incentive to work faster and better. Outworkers were also employed when needed; small workshops and family set-ups prepared such items as strips for borders and decorations, from which sections were cut to size in the factory.

It was Märklin's stock-keeping policies which really kept production economically viable. A vast stock was held of basic components, mechanisms, partially assembled items and semi-decorated toys, with the needs of one or two years catered for. The manufacture of dies, tooling and the mounting of tools was very expensive, so to keep costs down enormous runs of the different components were produced at one session and held in storage. Thus the basic parts for many toys could be taken from stock as the need arose, without having to re-mount the machinery. Decorative borders and embossed façades and roofs were usually made up by the metre and portions cut off when necessary.

Further rationalization occurred in the wide use of the same components, such as wheels, in a great many different toys. By clever design and planning parts could be adapted or used in all sorts of unlikely ways. Imaginative cosmetic treatment gave a new character to many toys constructed from the same basic items; a standard locomotive of German outline could be 'Americanized' by the addition of a cow-catcher and bell, for example, or a station façade given a face-lift by alterations in windows, platform and notices. Thus the same basic die could be employed consistently over the years although its extensive re-use would not have been obvious to the customer.

Without such systems it would have been impossible for Märklin to offer in the years before the First World War such a wide range of seemingly different toys. These systems also stood the firm in good stead after the war, when Märklin was able to resume toy production almost immediately after 1918; basic components had been held in stock throughout the war, as had semi-finished items and a number of completed toys left stranded by orders cancelled in 1914. Thus the effective use of production stock-piling enabled Märklin to keep costs in control and profits flowing in.

Sales and Marketing

As Märklin's range expanded and the number of export markets the firm supplied increased, so more efficient and organized sales policies were needed. In the early years of the company's history most of the selling was undertaken by Caroline and subsequently other representatives from the firm; the market place was confined to Germany and neighbouring countries, particularly Switzerland. The representatives travelled with samples and with maquettes of new toys, taking orders from customers. Hand-coloured sample books helped to demonstrate the range on offer and clients were at liberty to request items in different colour schemes and sizes. In these years toys were mostly only made up when firm orders had been placed.

The early 1890s saw the arrival of Emil Friz at Märklin, the injection of his capital into the firm and the introduction of several influential developments (such as the first train set with a sectional track); consequently the firm's horizons began to expand beyond the home and local markets. It soon began to make up some of the ground gained by Bing and other competitors, who were already directing much of their business towards customers abroad. In order to cope with the growing sales in other countries, Märklin gradually introduced more sophisticated sales methods.

The Leipzig Fair was (until 1945) the main international selling forum in the toy world. All the major toymakers exhibited here each year, showing prototypes of new models, refined and updated versions of toys already in production and a good selection of their most popular lines. The fair was strictly for the trade, rather than the public. Rival manufacturers eyed each other's products and took note of new developments by their competitors – there was no patent system in the early years to prevent them copying one another, despite the fact that most companies trademarked their products.

Wholesalers, importers and other clients from all over the world congregated at Leipzig to weigh up the merits of different products and to compare prices and delivery times. Orders were placed for toys to be delivered in time for the next Christmas season. Few retailers attended the fair, since they did their buying through the wholesalers and importers, though Mr A. W. Gamage, the proprietor of Gamage's department store in London, was a notable exception. His special arrangements with the German toymakers meant that he attended the fair regularly; for many years he was Märklin's sole importer in Britain and held exclusive rights on a number of the firm's toys.

Märklin certainly exhibited at the fair consistently from 1891 and it is more likely that the firm's ranges were represented there prior to this date, even if not in the years when Caroline ran the company on her own. And it was probably here at Leipzig that Märklin first made contact with the foreign importers and wholesalers who were to prove so crucial for the firm's ultimate success. Märklin also showed at many other international and national fairs and exhibitions, winning a number of prizes and accolades. Such displays boosted the company's reputation and drew the attention of a wider audience to its wares.

After 1900 Märklin accelerated its efforts to capture a significant proportion of the lucrative export market; as we have seen, ranges were increased dramatically and toys were tailored to specific national markets and given appropriate shapes and names. Despite the progress already made in this field by the Nurembergers, Märklin was able to catch up and even outstrip other companies in the export business. This expansion was assisted by the flexible attitude maintained by the company towards its customers and their requests: Märklin was prepared to make special deals with clients, to produce exclusive lines for them, to accept their ideas for toys and to take note of their advice on saleable products.

International sales increased yet further with the arrival in 1907 of Richard Safft whose main purpose was to expand the export ranges and markets. By now most major countries bought large quantities of Märklin toys, the USA, Britain and France being among the most important customers. In America there were few national toymakers who could achieve the same quality and scale of production; in Britain it was cheaper to import from Germany than to buy items made at home. This latter market was of particular interest to Märklin; the relationship was cultivated on both sides and became mutually beneficial. It was in England that the model railway craze flourished and Märklin was quick to see the possibilities of supplying the needs of a growing number of enthusiasts. After the war the way in which the firm reacted to this new fashionable hobby enabled it to become the manufacturer of some of the finest model railways.

The redoubtable Mr Gamage was also instrumental in Märklin's success in Britain. Märklin made up special orders and lines for sale at Gamage's and even manufactured toys which bore the Gamage trademark – an unthinkable move in the eyes of many rival toymakers. On the standard Märklin toys sold to Gamage's the Märklin name was not pushed to the fore; though most of the toys featured a trademark somewhere, Märklin, unlike a number of other toymakers, did not insist on the firm's identity being proclaimed at every opportunity. When the international situation worsened in the years immediately prior to the outbreak of the First World War, Märklin set aside patriotic pride and made efforts to conceal the origins of its toys in a diplomatic move designed to ensure continued sales abroad for as long as possible. (The 1911/12 French catalogue gives only the Paris address of 'Maerklin Frères & Cie'.)

Märklin's strategy-conscious sales philosophy forced other makers to re-think their own ideas; some followed suit, others went their own way. Of those who took note, many began to sell almost exclusively abroad. However, Märklin, whether by design or circumstance, had never neglected the home market, so was not as damaged as a large number of other firms by the consequences of the war. The attention given to sales outlets at home meant that business with long-established customers in Germany could be recommenced almost immediately after the war, while waiting for the resumption of trade abroad. Other firms barely recovered from the damage done to their main export outlets by the war and, with the ensuing depression, found it almost impossible to generate the same kind of business afterwards.

In order to organize its export sales Märklin built up an efficient network of wholesalers, agents and importers abroad, many of whom visited Göppingen, attended Leipzig and were visited by Märklin's representatives. Before the First World War there was an agent in every main country of export. The agents usually acted as wholesalers for Märklin and often had exclusive rights on Märklin toys in their particular country. The retailers put in orders for Märklin toys to the agents, who probably represented other firms as well and displayed in their showrooms a range of toys from several German toymakers. A certain amount of stock was held by each agent, but large orders were processed through the factory. It is thought that representatives were also used, and known that special clients, such as Mr Gamage, who combined the roles of importer and retailer, made their own arrangements directly with the factory.

In both Paris and London, and almost certainly in the USA and other importing countries, the wholesaler/agent also ran a repair service on Märklin's behalf. This meant that toys could be repaired locally without the need for them to be returned to Göppingen. Models could be adapted or mechanisms exchanged there, if the toy required was not immediately available; if, for example, a retailer had had a request for a particular locomotive with a clockwork motor, but this model was not held in stock, a steam version could be adapted and fitted with a clockwork mechanism at the agent's workshop and swiftly delivered to the customer.

Though, as has been mentioned, the wholesaler/agent held in stock a good number of toys, his main sources of reference of all the possible permutations of different lines on offer were the comprehensive catalogues produced by Märklin. The early hand-coloured sample books were dispensed with as soon as the Märklin range really expanded and were superseded by comprehensive black and white printed catalogues. These vast tomes, in which almost every toy was illustrated, described and numbered, are thought to have been prepared at Märklin, though printed outside by a local Göppingen firm. From them the wholesalers/agents compiled their own catalogues, which probably also included selections of items from other firms they represented. Again the case of Mr Gamage was an exception; no intermediary wholesale catalogue was prepared, only the Gamage's retail store catalogue.

The Märklin catalogues were issued at relatively regular intervals, usually about every eighteen months to two years. Supplements showing the new models developed in the interim were supplied periodically between the main volumes. From 1904 the full range was reclassified; attempts were made to sub-divide the lines into subject categories and into separate export ranges. Each sub-division received one of the following codes, indicating its content: *H*, *J*, *L*, *M*, *O* and *Fidelio*. The choice of these specific codes seems rather eccentric, especially since the codes bore little apparent relation to the categories of toys they represented – *M* for railways, for example. Each code was followed by a number which indicated the catalogue's date of issue; thus *M9* was distributed in 1909, *M10* in 1910 and so on. These sub-divisions meant that it was not necessary for a wholesaler/agent to take the entire catalogue, but only the sections which were of interest to him.

The above description gives a highly simplified explanation of the workings of the Märklin catalogues. As collectors familiar with these volumes know, the classification is somewhat confused and does not consistently follow one system or the other. This seems particularly strange considering Märklin's efficiency in other areas. By contrast, the equivalent volumes offered by other manufacturers, notably Bing, are much more systematically organized. However, the Märklin catalogues, and the illustrations and descriptions they contain, are one of our main sources of knowledge today about the ranges, progress and workings of the Märklin business; a study of them provides a wealth of fascinating information and it is easy to see why collectors interested in historical detail constantly refer to them.

Apart from giving information on dates, prices and methods of manufacture, the preliminary pages of some of the catalogue supplements prepared for the export markets make interesting reading. The names and addresses of the agents in several countries are listed and showrooms in Berlin, Hamburg, Paris, London (interestingly, not Gamage's), Amsterdam, Milan and Madrid mentioned. Details are also sometimes given of the far-flung corners of the world to which Märklin toys could be despatched; in one French supplement these include Persia, Siam, Indo-China, New Caledonia, Mozambique and the Congo. That the example of the Wright brothers biplane shown on pages 156-7 was unearthed in a palace in Thailand (Siam) comes as less of a surprise after learning from the catalogues that a Märklin toy might arrive as a Christmas present almost anywhere in the world – some indication of the lengths to which Märklin was prepared to go in the distribution of its wares.

Research still remains to be done into the history and workings of the Märklin business in order to complete our knowledge of the fascinating period leading up to the First World War. In the meantime this introduction may serve to underline and to some extent explain the unparalleled achievements and importance of the late 19th- and early 20th-century products of a toymaking firm which (appropriately) gave its motto in the 1900-2 catalogue as:

'Vom Guten das Beste!'
(Of the good, the best!)

Charlotte Parry-Crooke
London, 1982

Catalogue Reference Index

This index provides factual details on all the toys shown individually in this book and most of those illustrated in the group photographs.

Specific Märklin catalogue references are given, where they exist.

Where objects illustrated do not appear in any catalogue, catalogue references of similar items are supplied whenever possible, and these references are marked with the symbol *. The Märklin catalogue code is followed by the object's catalogue reference number. Note that each catalogue had its own numbering system and that therefore the same item appearing in two or more catalogues will not necessarily have the same number.

Dimensions are rounded up or down when appropriate. In all instances objects on each page opening are listed from left to right.

Objects are in private collections unless specifically stated.

Index

Cet index fournit des informations détaillées sur tous jouets illustrés individuellement dans ce catalogue ainsi que sur la plupart de ceux qui apparaissent dans les photographies de groupe.

Les références spécifiques au catalogue Märklin sont données quand elles existent.

Quand les objets illustrés n'apparaissent dans aucun catalogue, les références fournies quand il est possible, marquées d'un *, sont celles d'objets similaires. Le code du catalogue Märklin est suivi du numéro de référence de l'objet. Notez bien que chaque catalogue a son propre système de numérotation et qu'il se peut donc que le même objet figurant dans deux catalogues ou plus n'ait pas nécessairement le même numero.

Les dimensions au besoin ont été arrondies. Dans tous les cas les objets indexés ci-dessous apparaissent de gauche à droite sur les illustrations.

Les objets sont dans des collections particulières à moins d'indication contraire.

Katalogverweise

Dieser Index enthält Detailangaben über alle in diesem Buch einzeln sowie über die meisten auf den Gruppenfotos abgebildeten Spielzeuge.

Soweit vorhanden werden die jeweiligen Märklin-Katalogbezeichnungen angegeben. Wenn im Buch abgebildete Objekte in keinem Katalog verzeichnet sind, werden, so weit wie möglich, die Katalogbezeichnungen ähnlicher Stücke angeführt und mit dem Symbol * gekennzeichnet. Auf die Märklin-Katalogbezeichnung folgt die Katalogeintragnummer des jeweiligen Objekts. Dabei ist zu beachten, daß jeder Katalog seine eigene Numerierung hatte und ein in zwei oder mehreren Katalogen aufgeführtes Spielzeug daher mehrere unterschiedliche Katalogeintragnummern haben kann.

Abmessungen wurden, sofern erforderlich, auf- bzw. abgerundet. In allen Fällen werden die auf jeder geöffneten Buchseite erscheinenden Stücke von links nach rechts aufgeführt. Sofern nicht anders angegeben, befinden sich die Objekte in Privatbesitz.

28 | 29
PRR Locomotive
H4. AB1020
11 x 25 cm [$4\frac{3}{8}$ x $9\frac{7}{8}$ in]
LNWR Locomotive
H4. E1020
12 x 34 cm [$4\frac{3}{4}$ x $13\frac{3}{8}$ in]
30 | 31
PLM Locomotive
CV. 1020
11 x 36 cm [$4\frac{3}{8}$ x $14\frac{1}{8}$ in]
Station / Gare / Bahnhof
L9. 2012
28 x 35 cm [11 x $13\frac{7}{8}$ in]
Footbridge
Passerelle
Fußgängerbrücke
L9. 2834/0
19 x 37 cm [$7\frac{1}{2}$ x $14\frac{5}{8}$ in]
32 | 33
L9. 2006
33 x 39 cm [13 x $15\frac{3}{8}$ in]
34 | 35
H4. 2140/1
30 x 43 cm [$11\frac{7}{8}$ x $16\frac{7}{8}$ in]
36 | 37
H4. 2133
22 x 27 cm [$8\frac{5}{8}$ x $10\frac{7}{8}$ in]
38 | 39
L9. 2017
48 x 66 cm [$18\frac{7}{8}$ x 2 ft 2 in]
40 | 41
H4. 2640/2
28 x 35 cm [11 x $13\frac{7}{8}$ in]
42 | 43
H4. 1825
26 x 34 cm [$10\frac{1}{4}$ x $13\frac{3}{8}$ in]
44 | 45
1895. 1024B
19 x 38 cm [$7\frac{1}{2}$ x 15 in]
London Toy & Model Museum
46
9 x 12 cm [$3\frac{1}{2}$ x $4\frac{3}{4}$ in]
47
L9. R1061
14 x 24 cm [$5\frac{1}{2}$ x $9\frac{1}{2}$ in]
48
H4. AR
14 x 34 cm [$5\frac{1}{2}$ x $13\frac{3}{8}$ in]
49
L9. R0
14 x 35 cm [$5\frac{1}{2}$ x $13\frac{7}{8}$ in]
50 | 51
Locomotive
H4. 1045/2
15 x 50 cm [$5\frac{7}{8}$ x $19\frac{1}{4}$ in]
Water pump
Pompe d'eau
Wasserpumpe
L9. 2561
30 x 22 cm ($11\frac{7}{8}$ x $8\frac{5}{8}$ in]
52 | 53
Locomotive
L9. 3020
11 x 28 cm [$4\frac{3}{8}$ x 11 in]
Platform / Quai / Bahnsteig
L9. 2069/0
27 x 28 cm [$10\frac{5}{8}$ x 11 in]
54 | 55
Locomotive
H4. 1848/2
16 x 57 cm [$6\frac{1}{4}$ x $22\frac{3}{8}$ in]
56 | 57
Locomotive
L9. R4021/1
17 x 43 cm [$6\frac{3}{4}$ x $16\frac{7}{8}$ in]
Circus caravan
Roulotte de cirque
Zirkuswagen
L9. 1878/1
16 x 22 cm [$6\frac{1}{4}$ x $8\frac{5}{8}$ in]
58 | 59
Carriage / Wagon / Wagen
Archive No 4 JM
14 x 32 cm [$5\frac{1}{2}$ x $12\frac{5}{8}$ in]
Locomotive
H4. E1021/MR
14 x 47 cm [$5\frac{1}{2}$ x $18\frac{1}{2}$ in]
60 | 61
Locomotive
L9. FE3021 EB
18 x 58 cm [$7\frac{1}{8}$ x $22\frac{7}{8}$ in]
Carriage / Wagon / Wagen
L9. 1841/1
14 x 32 cm [$5\frac{1}{2}$ x $12\frac{5}{8}$ in]
Switch tower
Tour de contrôle
Stellwerk
L9. 3739
28 x 27 cm [11 x $10\frac{5}{8}$ in]
62 | 63
Locomotive
H4. D1021 CHD
14 x 42 cm [$5\frac{1}{2}$ x $16\frac{1}{2}$ in]
Carriage / Wagon / Wagen
H4. 1856/1
12 x 24 cm [$4\frac{3}{4}$ x $9\frac{1}{2}$ in]
Weighbridge
Bascule à wagons
Brückenwaage
L9. 2065/0
15 x 18 cm [$5\frac{7}{8}$ x $7\frac{1}{8}$ in]
64 | 65
H4. K4021
16 x 39 cm [$6\frac{1}{4}$ x $15\frac{3}{8}$ in]
66
P14. CL3020
13 x 27 cm [$5\frac{1}{8}$ x $10\frac{5}{8}$ in]

67
P14. CL3121
16 x 33 cm [6¼ x 13 in]
68
Gamage's 1902-1906
(Published by Denys Ingram, London)
14 x 24 cm [5½ x 9½ in]
69
00. V3020
11 X 18 cm [4⅜ x 7⅛ in]
Gantry crane
Grug à chariots
Portalkran
L9. 2358
25 x 32 cm [9⅞ x 12⅝ in]
70 | 71
Locomotive (1)
L9. 3191
14 x 31 cm [5½ x 12¼ in]
Locomotive (0)
L9. 1190
11 x 21 cm [4⅜ x 8¼ in]
Platform / Quai / Bahnsteig
L9. 2060/53
18 x 36 cm [7½ x 14¼ in]
72 | 73
L9. 3171/1
13 x 22 cm [5⅛ x 8⅝ in]
74
L9. 1071/62
13 x 22 cm [5⅛ x 8⅝ in]
75
L9. 1071/42
13 x 22 cm [5⅛ x 8⅝ in]
76
13 x 32 cm [5⅛ x 12⅝ in]
77
H4. 3031
13 x 22 cm [5⅛ x 8⅝ in]
78 | 79
H4. 2330/2
L9. 1080
10 x 24 cm [4 x 9½ in]
80 | 81
Carriage / Wagon / Wagen
H4. 1851/1
13 x 20 cm [5⅛ x 7⅞ in]
Locomotive
H4. K1021
15 x 33 cm [5⅞ x 13 in]
82 | 83
L9. SK1021
12 x 23 cm [4¾ x 9 in]
84 | 85
P10. 1914/1
14 x 32 cm [5⅝ x 12⅝ in]
86
L9. 1828/1
13 x 22 cm [5⅛ x 8⅝ in]
87
L9. 1848/1
14 x 28 cm [5½ x 11 in]
88 | 89
L9. 1888/0
8 x 16 cm [3¼ x 6¼ in]
L9. 1888/1
11 x 20 cm [4⅜ x 7⅞ in]
90 | 91
12 x 18 cm [4¾ x 7⅛ in]
92
9 x 13 cm [3½ x 5⅛ in]
93
L9. 2865/1
12 x 22 cm [4¾ x 8⅝ in]
94 | 95
L9. 2993/1
13 x 22 cm [5⅛ x 8⅝ in]
96
L9. 1909/1
13 x 22 cm [5⅛ x 8⅝ in]
97
9 x 12 cm [3½ x 4¾ in]
98 | 99
'Kaiser' coach (2)
Wagon du 'Kaiser' (2)
Kaiserzugwagen (2)
18 x 38 cm [7⅛ x 15 in]
'Kaiser' coach (1)
L9. 1847/1
14 x 32 cm [5½ x 12⅝ in]
'Kaiser' coach (0)
9 x 19 cm [3½ x 7½ in]
100 | 101
L9. 1824/1
12 x 22 cm [4¾ x 8⅝ in]
102 | 103
P10. 2991/1
16 x 23 cm [6¼ x 9 in]
104 | 105
L9. 1100/0
13 x 18 cm [5⅛ x 7⅛ in]
106 | 107
Inspection car / Draisine
L9. 1100/1
16 x 23 cm [6¼ x 9 in]
Platform roof
Verrière de quai
Bahnsteigdach
L9. 2067/0
23 x 29 cm [9 x 11½ in]
108
L9. 1877/0
9 x 14 cm [3½ x 5½ in]
109
L9. 1878
9 x 15 cm [3½ x 5⅞ in]
110 | 111
Mobile kitchen
Cuisine mobile
Feldküche
M9. 8035/1B
10 x 16 cm [4 x 6¼ in]
Transporter
Wagon de transport
Güterwagen
L9. 1818/2
8 x 24 cm [3¼ x 9½ in]
112 | 113
Locomotive
P10. CV1020
11 x 36 cm [4⅜ x 14⅛ in]
Carriage / Wagon / Wagen
P10. 1841/0
10 x 18 cm [4 x 7⅛ in]
Bridge / Pont / Brücke
L9. 1504/0
30 x 158 cm [11⅞ in x 5 ft 2¼ in]
114 | 115
L9. 2435
24 cm [17⅞ in]
L9. 35552/2
47 cm [18½ in]
116 | 117
L9. 2570
13 x 27 cm [5⅛ x 10⅝ in]
L9. 2685G
10 x 13 cm [4 x 5⅛ in]
London Toy & Model Museum
L9. 2597
9 x 7 cm [3½ x 2¾ in]
120 | 121
L9. 5210
18 x 30 cm [7⅛ x 11⅞ in]
122 | 123
L9. 5210
18 x 30 cm [7⅛ x 11⅞ in]
124 | 125
Car / Automobile / Auto
L9. 5214/3
25 x 46 cm [9⅞ x 18⅛ in]
Lamp / Lanterne / Laterne
L9. 3564
43 cm [16⅞ in]
126 | 127
L9. 5210
16 x 22 cm [6¼ x 8⅝ in]
Reverend Bryan Apps
128 | 129
L9. 5206
16 x 34 cm [6¼ x 13⅜ in]
London Toy & Model Museum
132 | 133
H. 1904 *
39 x 50 cm [15⅜ x 19¾ in]
London Toy & Model Museum
134 | 135
L9. 5064
52 x 72 cm [15 x 29⅛ in]
London Toy & Model Museum
136 | 137
L9. 5031
26 x 31 cm [10¼ x 12½ in]
138 | 139
L9. 5053 *
16 x 46 cm [6¼ x 18⅛ in]
140 | 141
L9. 4093/5121
52 x 72 cm [20½ in x 2 ft 4⅜ in]
London Toy & Model Museum
142 | 143
L9. 5080/75
L9. 4080/3
29 x 75 cm [11½ in x 2 ft 5½ in]
London Toy & Model Museum
144 | 145
H. 1904/1096
66 x 105 cm
[2 ft 2in x 2 ft 5½ in]
London Toy & Model Museum
146 | 147
H4. 1090
43 x 54 cm [16⅞ x 21¼ in]
148 | 149
L9. 5102
27 x 35 cm [10⅝ x 13⅜ in]
150 | 151
L9. 5050/7
43 x 72 cm [16⅞ in x 2ft 4⅜ in]
London Toy & Model Museum
152 | 153
L9. 5050/9D
53 x 98 cm
[20⅞ in x 3 ft 2⅝ in]
London Toy & Model Museum
156 | 157
L9. 5418
44 x 40 cm [17⅜ x 15¾ in]
London Toy & Model Museum
158 | 159
L9. 5403
12 x 38 cm [4¾ x 15 in]
160 | 161
M9. 8519
25 x 23 cm [9⅞ x 9 in]
London Toy & Model Museum
162 | 163
M9. 8512
21 x 27 cm [8¼ x 10⅝ in]
London Toy & Model Museum
164 | 165
M9. 8511
17 x 18 cm [6¾ x 7⅛ in]
166 | 167
M9. 8262
17 x 38 cm [6¾ x 15 in]
London Toy & Model Museum
168 | 169
M9. 84521B
19 x 47 cm [7½ x 18½ in]
London Toy & Model Museum
170 | 171
M9. 8847
54 x 45 cm [21¼ x 17¾ in]
London Toy & Model Museum
172 | 173
M9. 8965W
16 x 20 cm [6¼ x 7⅞ in]
London Toy & Model Museum
174
Monsieur G. Malafosse, Paris
175
L9. E4020
14 x 40 cm [5½ x 15¾ in]
176 | 177
M9. 8103
20 x 11 cm [7⅞ x 4⅜ in]
178
M9. 9694K *
30 x 32 cm [11⅞ x 12⅝ in]
179
M9. 9645N *
M9. 9673N *
43 x 20 cm [16⅞ x 7⅞ in]

RAILWAYS
TRAINS
EISENBAHNEN

In the foreground of this fine railway group is a 1904 clockwork PRR (Pennsylvania Rail Road) locomotive for gauge 0. Behind it is a 1905 LNWR (London & North Western Railway) clockwork locomotive, also for gauge 0; this pulls a 1907 ventilated fruit truck and a poultry truck. The station, made between 1906 and 1919, is American in style despite notices in German.

Ce bel ensemble comprend au premier plan une locomotive mécanique PRR (Pennsylvania Rail Road) en écartement 0. Derrière, une locomotive LNWR (London & North Western Railway) de 1905, mécanique également, en écartement 0, entraine deux wagons de transport, l'un de fruits datant de 1907, l'autre de volailles. La gare, fabriquée entre 1906 et 1919, est de style américain bien que les inscriptions soient en Allemand.

Den Vordergrund dieser hübschen Eisenbahngruppe bildet eine PRR (Pennsylvania Rail Road) Uhrwerklok von 1904, und hinter ihr steht eine LNWR (London & North Western Railway) Uhrwerklok von 1905, beide in Spur 0. Sie zieht einen belüfteten Obstwagen von 1907 und einen Geflügelwagen. Der zwischen 1906 und 1919 gebaute Bahnhof folgt trotz deutscher Beschriftung dem amerikanischem Stil.

L.N.W.R.
P.R.R.
90309
P.R.R.
P.R.R.

One of the first attempts at locomotive streamlining was developed in toy form in Märklin's famous PLM (Paris-Lyon-Meditérranée) wind-cutter. Between 1906 and 1919 four models were produced. This 1907 clockwork example for gauge 0 is the second and the most finely finished. The rare PLM mail van was manufactured between 1908 and 1914 and the footbridge in 1908.

La célèbre Coupe-Vent PLM (Paris-Lyon-Méditérranée) de Märklin est l'une des premières expériences d'aérodynamique qui ait été tentée sur une locomotive. Quatre modèles furent produits entre 1904 et 1907. Ce modèle mécanique de 1907 en écartement 0 est le deuxième et le plus fini. Le wagon poste PLM, peu courant, fut fabriqué entre 1908 et 1914 et la passerelle en 1908.

Eine der ersten stromlinienförmigen Loks war die berühmte PLM (Paris-Lyon-Meditérranée) Spielzeuglok von Märklin. Zwischen 1906 und 1919 wurden vier Modelle hergestellt. Diese Uhrwerklok, Spurweite 0, von 1907 ist das zweite und bestausgeführte Modell. Der seltene PLM Postwagen wurde zwischen 1908 und 1914, die Fußgängerbrücke 1908 hergestellt.

NOGENT
PASSAGE
CONTRÔLE
SALLE D'ATTENTE
Lyon
G.V.1020

Made in 1908, this splendid station shows the use of the first telegraph system which was initially installed to run in conjunction with the railway networks. The design of the station building is comparatively simple, apart from the strongly curved gables that are reminiscent of the northern European architectural style of the day. Lighting is provided by a lit candle placed inside the waiting-room. Another version of this station, with different windows and inferior paintwork, was put into production after the First World War in about 1920.

Fabriquée en 1908, cette splendide gare illustre l'emploi du premier télégraphe, installé à l'origine pour fonctionner de concert avec les réseaux ferroviaires. Le style de la gare est relativement simple à l'exception des pignons fortement arrondis qui évoquent l'architecture alors en vigueur dans le nord de l'Europe. Une bougie dans la salle d'attente procure de la lumière. Une seconde version de cette gare, avec des fenêtres différentes et une peinture de moindre qualité, fut produite après la première Guerre Mondiale, aux alentours de 1920.

Dieser gefällige Bahnhof von 1908 zeigt den erstmaligen Einsatz eines Telegraphensystems, das ursprünglich für den gemeinsamen Betrieb mit der Bahn installiert wurde. Abgesehen von den stark abgerundeten Giebeln – typisch für die damalige nordeuropäische Architektur – ist der Bahnhof relativ schlicht gebaut. Eine Kerze im Warteraum lieferte die Beleuchtung. Eine Abwandlung dieses Bahnhofs mit anderen Fenstern und weniger guter Lackierung wurde um 1920 hergestellt.

PASSAGE
SALLE D'ATTENTE
CHEF DE GARE
Bagage
TÉLÉGRAPHIE

This interesting station, featuring the early telegraph system, was the first of a series made from 1902 to 1909. It was marketed internationally, with appropriate language alterations for the notices. With its original central dome and arched arcades, it seems to have been influenced by the central European buildings of Byzantine inspiration. Subsequent models were produced in red and yellow.

Cette curieuse gare avec son télégraphe fut la première d'une série fabriquée entre 1902 et 1909. Elle fut vendue internationalement avec des inscriptions différentes suivant les pays. Avec son dome central et ses arcades cintrées elle semble avoir été inspirée par l'architecture d'expression Byzantine que l'on trouve en Europe centrale. Les modèles suivants furent produits en rouge et jaune.

Dieser interessante Bahnhof war zusammen mit der ersten Telegraphenanlage der erste einer von 1902 bis 1909 gebauten Reihe. Er wurde, versehen mit den entsprechenden fremdsprachigen Aufschriften, international vertrieben. Die Kuppel und die Bogeneingänge scheinen auf mitteleuropäische Bauwerke mit byzantinischem Einschlag zurückzugehen. Später hergestellte Modelle waren rot und gelb.

Central-Bahnhof
DURCHGANG
Vorstand
Gepäck
Wartesaal
Telegraph

This charming customs house is a rare model made for gauge 0 and gauge 1 layouts. It was offered in Märklin's catalogues between 1904 and 1909. The border, rail and finials on the roof, and the porches, with their embossed decorations and shaped pillars, are delightful features of a toy that exhibits a high quality of craftsmanship.

Ce charmant poste de douane est un modèle rare en écartement 0 et 1. On le trouvait dans les catalogues Märklin entre 1904 et 1909. Les épis et la ballustrade sur le toit ainsi que les porches avec leurs piliers ouvragés et décorations estampées caractérisent de façon charmante un jouet qui témoigne d'une haute qualité de fabrication.

Dieses niedliche Zollhaus ist ein seltenes, für die Spurweiten 0 und 1 hergestelltes Modell. Es wurde in den Märklin-Katalogen von 1904 bis 1909 angeboten. Die Spitzen, die Stange und die Blumenverzierungen auf dem Dach und die Eingänge mit ihren Prägungen und profilierten Säulen sind hübsche Attribute eines hochwertigen Spielzeugs.

Douane
Douane
Entrée
Sortie

Though this is not the rarest Märklin station, it is one which illustrates excellently the architecture of the 1900s, iron successfully blending with stone. The manufacture of the brickwork façade required extensive pressing and embossing. Available for layouts in both gauge 0 and 1, it was marketed from 1904 to 1909. The manually operated bell is shown above.

Bien qu'elle ne soit pas la plus rare des gares Märklin, elle illustre de façon excellente l'architecture des années 1900 qui mariait avec succès le fer et la pierre. Disponible en écartement 0 et 1, elle fut vendue de 1904 à 1909. On peut voir ci-dessus la cloche que l'on faisait fonctionner manuellement.

Dies ist zwar nicht der seltenste Märklin-Bahnhof, wohl aber einer, der den Baustil um 1900 und die gelungene Kombination von Eisen und Stein gekonnt wiedergibt. Die Fassade wurde mit einem hohen Aufwand an Preß- und Prägearbeiten hergestellt. Der Bahnhof für die Spurweiten 0 und 1 wurde zwischen 1904 und 1909 gebaut. Oben ist die von Hand betätigte Glocke abgebildet.

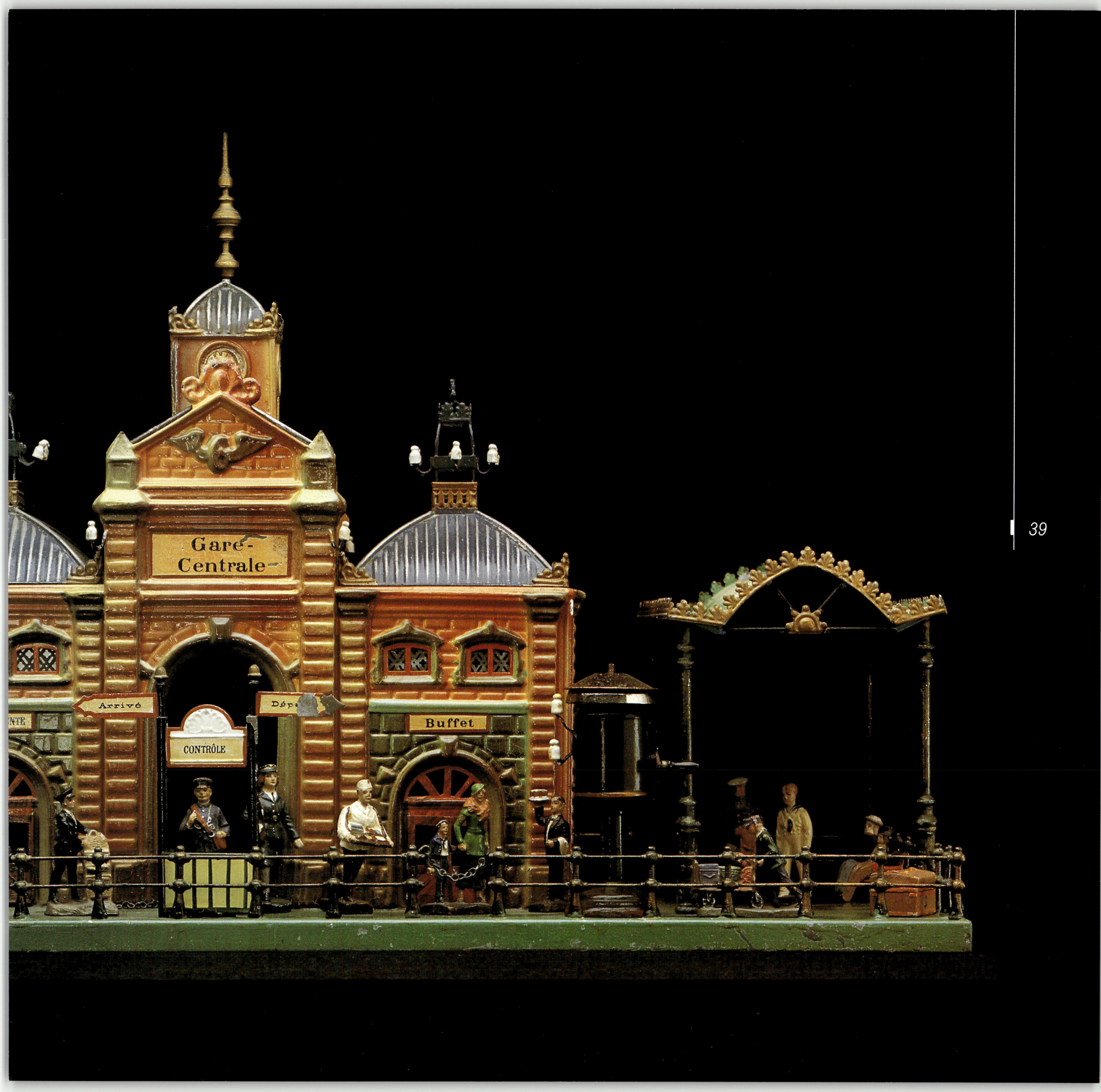
Gare-
Centrale
Arrivé
CONTRÔLE
Buffet

The design of this ornate station, with its mansard roof, shows the influence of French buildings of the 1880s, such as the Gare d'Orsay, Paris. Made for gauge 0 layouts, it sold from 1904 to 1919. The lead and iron handrail was replaced by a tin rail on the 1925 model. The station master in his control box is shown above.

Le style de cette gare très ouvragée avec son toit Mansard, témoigne de l'influence des ouvrages Français des années 1880 telle la Gare d'Orsay à Paris. Elle fut vendue de 1904 à 1919 en écartement 0 et 1. La rambarde en plomb et fer fut remplacée sur le modèle de 1925 par une en fer-blanc. Le chef de gare est dans son poste de contrôle.

Dieser reich verzierte Bahnhof mit seinem Mansardendach zeigt den Einfluß der französischen Baukunst um 1880, insbesondere des Gare d'Orsay in Paris. Der für Anlagen der Spur 0 hergestellte Bahnhof wurde zwischen 1904 und 1919 vertrieben. 1925 wurde der Blei- und Eisenhandlauf durch ein Zinngeländer ersetzt. Oben: Der Fahrkartenkontrolleur an der Sperre.

NOGENT
PASSAGE
CONTRÔLE
SALLE D'ATTENTE
SALLE D'ATTENTE
TÉLÉGRAPHIE

Made between 1906 and 1919, this toy is the epitome of a small country station. Several versions of this quite simply designed model were available for the different national markets.
The clock face below was rubber stamped onto the surface and then varnished to seal the design.

Ce jouet, fabriquée entre 1906 et 1919, est l'exemple même de la petite gare de campagne. Plusieurs versions de ce modèle, au style simple, étaient disponibles suivant les différents marchés nationaux.
L'horloge que l'on voit ici était imprimée au tampon puis vernie pour fixer le motif.

Dieses von 1906 bis 1919 mit einfachen Mitteln hergestellte Modell einer ländlichen Bahnstation war in mehreren Ausführungen für ausländische Märkte erhältlich.
Das unten gezeigte Zifferblatt wurde aufgestempelt und dann mit farblosem Lack versiegelt.

DURCHGANG
artesaal
Telegraph

This fine early gauge 3 clockwork train set was offered with engine, tender, two passenger coaches, a postal van and a baggage van. It was made from about 1895 and is similar to some trains produced by the R&GN, supporting the theory that Märklin took over some of the R&GN tools. It was one of the first groups to run on rails and was certainly produced before electric and possibly steam trains were available.

Ce beau train mécanique en écartement 3 était vendu complet avec locomotive, tender, deux wagons de passagers, un wagon poste et un wagon à bagages. Il fut fabriqué aux alentours de 1895, et ressemble à certains des trains fabriqués par R&GN, un point en faveur de la théorie selon laquelle Märklin aurait repris certains des outils de R&GN. Ce fut l'un des premiers ensembles à rouler sur rails et il fut certainement produit avant que les trains électriques et peut-être même à vapeur n'aient été disponibles.

Dieser hübsche frühe Uhrwerkzug, Spurweite 3, wurde mit Lok, Tender, zwei Passagier-, einem Post- und einem Gepäckwagen angeboten. Der ab etwa 1895 hergestellte Zug ähnelt einigen der Züge von R&GN und unterstützt daher die Ansicht, daß Märklin einige Werkzeuge und Vorrichtungen von R&GN übernommen hatte. Der Zug lief als einer der ersten auf Schienen und wurde bereits gebaut, noch ehe Elektro- und wahrscheinlich auch Dampfzüge erhältlich waren.

Made for a steep incline railway, this rare 1908 gauge 1 rack train was produced with variations from 1902 to 1919. Both engine and carriage were modelled on the Swiss and German mountain railways; a cogwheel on the locomotive engages with a continuous rack fixed between the rails. This version of the carriage does not appear in any catalogue.

Fabriqué pour les lignes à fortes pentes, ce rare train à crémaillère de 1908 en écartement 1 fut produit avec de légères variantes de 1902 à 1919. Locomotive et wagon furent tous deux copiés sur les trains de montagne allemands et suisses; une roue d'engrenage sous la locomotive s'enclanche sur la crémaillère fixée entre les rails. Il n'y a nulle trace de ce wagon dans les catalogues.

Diese seltene, für starke Steigungen mit Spurweite 1 bestimmte Zahnradbahn aus dem Jahre 1908 wurde von 1902 bis 1919 in mehreren Abwandlungen gebaut. Lok und Wagen folgen dem Vorbild der schweizerischen und deutschen Bergbahnen: ein Zahnrad unter der Lok greift in die Zahnstange ein. Diese seltene Version steht in keinem Katalog.

Shown below is a 1905 steam RO in gauge 0 made for America. Special modifications on the model for this market included cow-catcher, bell, high chimney and three rectangular windows.

Les modifications spéciales sur cette RO à vapeur de 1905 en écartement 0 pour l'Amérique comprennent un chasse-boeufs, une cloche, une haute cheminée et trois fenêtres rectangulaires.

Das Bild unten zeigt eine für Amerika in Spurweite 0 gebaute Dampflok mit Kuhfänger, Glocke, hohem Schornstein und drei rechtwinkligen Fenstern als Sonderausstattung für diesen Markt.

One of the first steam locomotives is shown on the right; this example, for gauge 0, was issued in 1902. A wooden side rail on each side prevents fingers from burning.

A droite, l'une des premières locomotives à vapeur; ce modèle en écartement 0 fut produit en 1902. Un support en bois de chaque côté de la locomotive empêche que l'on ne se brûle les doigts.

Rechts eine der ersten Dampfloks. Dieses Modell für Spurweite 0 kam 1902 heraus. Holzstäbe auf beiden Seiten schützen die Finger vor Verbrennungen.

An interesting goods depot, made from 1904 to 1909, is illustrated here with the PLM wind-cutter. The unusual water pump on the far right is manually operated and equipped with a lamp. The detail below includes a selection from the wide range of accessories that Märklin offered for use with the railway sets.

Un curieux entrepôt de marchandises, fabriqué de 1904 à 1909, est illustré ici avec la Coupe-Vent PLM. A droite, la pompe à eau est opérée manuellement et equipée d'une lampe. Le détail ci-dessous comprend une sélection d'accessoires tirée de l'importante gamme que Märklin offrait de pair avec ses trains.

Der interessante, zwischen 1904 und 1909 gebaute Güterbahnhof ist hier mit der PLM-Stromlinienlok abgebildet. Die ungewöhnliche Wasserpumpe ganz rechts wird von Hand bedient und besitzt eine Laterne. Der Ausschnitt hier zeigt eine Auswahl des vielfältigen Zubehörs, das Märklin für die Bahnen anbot.

C.V. 1021.
P.L.M.

This 1908 electric gauge 0 locomotive was also available in clockwork form. The carriages (for the PLM wind-cutter) were made from 1906 to 1912. The patterned glass roof of the covered platform was engraved rather than printed. The platform's design recalls the cast-iron and glass structures of the French architect, Baltard, who was responsible for the old Les Halles market in Paris. The building below is a typical Märklin accessory.

Cette locomotive électrique en écartement 0 était aussi disponible en modèle mécanique. Les wagons (pour la Coupe-Vent PLM) furent fabriqués entre 1906 et 1912. Le motif sur le toit de verre du quai couvert n'était pas imprimé mais gravé. Le style du quai n'est pas sans évoquer les structures de verre et métal de Baltard, le constructeur des Halles. L'édifice ci-dessous est caractéristique des accessoires Märklin.

Diese elektr. Lok von 1908, Spur 0, wurde auch als Uhrwerkversion angeboten. Die Wagen (für die PLM-Stromlinienlok) wurden von 1906 bis 1912 hergestellt. Das gläserne gemusterte Bahnsteigdach wurde geprägt, nicht bedruckt. Der Bahnhof erinnert an die Gußeisen- und Glaskonstruktionen des französischen Architekten Baltard, der den alten Großmarkt Les Halles in Paris schuf. Das Bauwerk unten ist ein typisches Märklin-Zubehör.

A typical product for the American market, this locomotive was made in about 1907. Despite its date, and the addition of cow-catcher and bell, it was based on the European outline. As the detail shows, the interior of the accompanying hospital van has many beautifully reproduced realistic features; these include bunks, stretchers, injured people, nurses, a cupboard for dressings and a stove. This carriage is certainly a perfect example of Märklin's attention to detail.

Cette locomotive fut fabriquée vers 1907 pour le marché américain. Malgré sa date et l'addition du chasse-boeufs et de la cloche, elle reposait sur le modèle européen. Comme l'on peut voir sur le détail, l'intérieur du wagon croix-rouge est remarquablement amménagé et ce de façon très réaliste: il y a couchettes, brancards, blessés, infirmières, un placard à pansements et un poële. Ce wagon illustre bien l'attention que Märklin apportait au détail.

Diese um 1907 hergestellte Lok war ein typisches Erzeugnis für den amerikanischen Markt. Trotz Baujahr und Hinzufügung von Kuhfänger und Glocke stützte sie sich auf europäische Vorbilder. Der Sanitätswagen ist innen wirklichkeitsgetreu ausgestattet: Kojen, Bahren, Verwundete, Schwestern, Verbandzeugschrank und Ofen. Dieser Wagen ist ein hervorragendes Beispiel für die Detailtreue Märklins.

This 1906 live steam engine, R1, was probably based on a design by the Austrian engineer, Gölsdorf, who was well-known for his work with the Austrian State Railways.

Cette locomotive Rl de 1906 à vapeur vive fut probablement imitée d'un plan de l'ingénieur autrichien Gölsdorf, célèbre pour ses travaux pour les chemins de fers autrichiens.

Die Dampflok R1 aus dem Jahre 1906 geht wahrscheinlich auf einen Entwurf des österreichischen Konstrukteurs Gölsdorf zurück, bekannt durch seine Arbeit für die österreichischen Bahnen.

Two wagons accompany the R1: a 1904 furniture truck and a 1908 circus caravan; both are on transporters which could be bought separately. Versions with alternative lettering were marketed abroad.

Deux wagons accompagnent la R1: un transport de meubles de 1904 et une roulotte de cirque; les bases de wagons pouvaient être achetées séparément.

Die R1 zieht zwei Güterwagen mit einem Möbelwagen (1904) und einem Zirkuswagen (1908), die separat gekauft werden konnten. Für das Ausland wurden die Wagen mit anderen Beschriftungen versehen.

Specially constructed in English, as opposed to European, outline for the British market, this interesting Midland Railway group was distributed by Gamage's of London. The clockwork gauge 1 locomotive dates from 1904 and bears the railway company's maroon livery. It pulls two fine carriages: a passenger car and a dining car. The standard lamp is of the same period.

Construit spécialement sur un profil anglais pour le marché britannique, ce curieux ensemble Midland Railway était distribué par le grand magasin Gamage à Londres. La locomotive mécanique en écartement 1 date de 1904 et porte la couleur bordeau de la compagnie. Elle entraine deux beaux wagons: un wagon de voyageurs et un wagon restaurant. Le lampadaire est de la même époque.

Diese interessante Midland Railway-Gruppe wurde für den britischen Markt nach englischen Vorlagen hergestellt und in London von Gamage's verkauft. Die Uhrwerklok, Spurweite 1, stammt aus dem Jahre 1904 und ist in den braunen Farben der Eisenbahngesellschaft gehalten. Die Lok zieht zwei hübsche Midland Railway-Wagen: einen Personenwagen und einen Speisewagen. Auch die Laterne stammt aus der gleichen Zeit.

This electric FE was made in 1907 for gauge 1. It is coupled here to four 'Kaiser' coaches: a sleeper, a passenger car, a saloon and a dining car; the rarest is the third, the saloon. The crown is the emblem of the Kaiser.

All four carriages were produced in many variations for each of the different export markets, including LNWR and GWR (Great Western Railway) liveries for England.

Cette FE électrique en écartement 1 fut fabriquée en 1907. Elle est accrochée ici aux quatre 'wagons du Kaiser': le wagon-lit, le wagon de voyageurs, la voiture-salon et le wagon-restaurant; la voiture-salon est le plus rare des wagons. Les quatre wagons furent produits pour l'exportation avec de nombreuses variantes y compris les couleurs des compagnies LNWR et GWR (Great Western Railway) pour le marché anglais.

Diese elektrische FE wurde 1907 für die Spur 1 hergestellt. Hier zieht sie vier Wagen des „Kaiserzuges", einen Schlafwagen, einen Personenwagen, einen sehr seltenen Salonwagen und einen Speisewagen.

Die Krone ist das Symbol des Kaisers. Die vier Wagen wurden in vielen Ausführungen für den Export hergestellt, darunter auch in den Farben der LNWR und der GWR (Great Western Railway) in England.

The famous 'Charles Dickens' locomotive was made for the English market between 1904 and 1907. It was offered in both electric and clockwork form in gauges 0, 1 and 2; the rarest is the gauge 0 model. The LNWR gauge 1 example illustrated pulls two LNWR carriages; different versions were produced for France and Germany. On the left is a weighbridge, available between 1902 and 1912.

La célèbre locomotive 'Charles Dickens' fut fabriquée pour le marché anglais entre 1904 et 1907. Elle existait en modèles électrique ou mécanique, en écartement 0,1 et 2; le plus rare est celui en écartement 0. Le modèle LNWR en écartement 1 illustré ici entraine deux wagons de la même compagnie; différentes versions furent produites suivant les pays. A gauche, une bascule à wagons disponible de 1902 à 1912.

Die berühmte „Charles Dickens“ wurde zwischen 1904 und 1907 für den englischen Markt hergestellt. Sie wurde als elektr. und Uhrwerklok für die Spuren 0, 1 und 2 angeboten, von denen das Modell 0 am seltensten ist.

Die hier abgebildete LNWR-Ausführung, Spur 1, zieht zwei LNWR-Wagen. Andere Ausführungen wurden für Frankreich und Deutschland gefertigt. Links eine zwischen 1902 und 1912 angebotene Brückenwaage.

This significant locomotive, called a 'Kassel-Hannover', was made in 1904 for gauge 1 only. The original prototype locomotive, designed by the engineer Henschel, had a large tender; this was omitted in Märklin's version in order to avoid excessive swing on running. The original was an experiment; only three engines were made, each with its driver up front and fireman at the back of the boiler. Representing one of the earliest attempts at streamlining, this toy, and its original counterpart, are worthy of the futuristic machines of Jules Verne's universe.

Cette impressionnante locomotive, la 'Kassel-Hannover', fut fabriquée en 1904 en écartement 1 uniquement. Le prototype original conçu par l'ingénieur Henschel avait un grand tender; ce trait fut omis dans la version Märklin afin d'éviter un trop grand balancement pendant la marche. L'original était expérimental: trois locomotives seulement furent construites, chacune avec conducteur debout à l'avant et chauffeur derrière la chaudière. Représentant l'un des premiers essais d'aérodynamique, ce jouet, ainsi que son modèle original, évoquent sans conteste les machines futuristes de l'univers de Jules Verne.

Diese „Kassel-Hannover" genannte auffällige Lok wurde 1904 nur in der Spur 1 gebaut. Das von der Firma Henschel gebaute echte Vorbild hatte einen großen Tender, der in der Märklin-Ausführung zur Vermeidung zu starker Schwingungen während der Fahrt weggelassen wurde. Das Original war eine in nur zwei Exemplaren gebaute Versuchslok, in der der Lokführer vor und der Heizer hinter dem Kessel standen. Die Spielzeuglok und ihr Vorbild sind erste Versuche einer stromlinienförmigen Gestaltung und erinnern an die futuristischen Maschinenschöpfungen eines Jules Verne.

17
R. H. V

This 1910 electric PO (Paris-Orléans) steeple-cab for gauge 0 is nicknamed *Boîte-à-Sel*. It was based on the Central London Railway engine.

Cette motrice électrique PO (Paris-Orléans) de 1910 en écartement 0 est surnommée 'Boite -à-Sel'. Elle était copiée sur la locomotive Central London Railway.

Diese elektrische PO-Lok (Paris-Orléans) von 1910 (Spur 0) trug den Spitznamen *Boîte-à-Sel*. Ihr Vorbild war die Central London Railway-Lok.

Two electric motors create great tractive power in this gauge 1 PO steeple-cab of 1908. The original locomotive on which this toy is based ran between Paris and Orléans.

Cette motrice PO en écartement 1 développait une grande force de traction grâce à ses deux moteurs électriques. L'original sur lequel ce jouet se base faisait le trajet Paris-Orléans.

Die hohe Zugleistung dieser PO-Lok von 1908 (Spur 1) kommt von zwei Elektromotoren. Das echte Vorbild war eine zwischen Paris und Orléans laufende Lok.

The 1906 clockwork gauge 1 Central London Railway 'No 23' is based on a locomotive used in a marshalling yard. The original was also the first locomotive on the English 'Twopenny Tube'.

Cette petite Central London Railway 'No 23' mécanique en écartement 1 est basée sur une locomotive utilisée dans les gares de triage. L'original fut aussi la première locomotive du métro anglais.

Das Vorbild dieser „No. 23", einer Central London Railway-Uhrwerklok von 1906 mit Spurweite 1, war eine Rangierlok, die als eine der ersten Loks auf der englischen U-Bahn benutzt wurde.

This 1906 four-wheeled electric version of the Central London Railway 'No 23' was superseded by an eight-wheeled model. The gantry crane was sold between 1904 and 1912.

Cette version électrique à quatre roues de la Central London Railway 'No 23' fut remplacée par un modèle à huit roues.
La grue à chariot fut vendue entre 1904 et 1912.

Diese zweiachsige elektrische Variante der „No. 23" der Central London Railway von 1906 wurde durch ein vierachsiges Modell ersetzt. Der Portalkran wurde zwischen 1904 und 1912 verkauft.

The accessories shown with the two versions of the Paris Métro recall French railway architecture of the time; the reproductions of enamel advertisements give an appropriate period atmosphere.

Les accessoires représentés ici avec deux versions du métro parisien rapellent l'architecture ferroviaire française de l'époque. Les publicités en fer émaillé rendent l'atmosphère de cette période.

Das Zubehör zu den beiden Ausführungen der Pariser Metro zeigt die zeitgenössische französische Eisenbahnarchitektur. Die emaillierten Reklameschilder vermitteln die Atmosphäre jener Zeit.

Two rare versions of the Paris Métro train are shown here: a clockwork gauge 0 example of 1907 and an electric gauge 1 example of 1906. The conductor shoe on the electric motor coach is designed to pick up the current from an outside conductor rail – an unusual feature on a toy. The Métro was probably destined for the French market only.

Ces deux versions du métro parisien sont rares: l'une mécanique en écartement 0 date de 1907, l'autre électrique en écartement 1 de 1906. Dans le modèle électrique la prise de courant est conçue pour capter le courant qui passe dans un rail extérieur – détail peu commun dans un jouet. Le métro était probablement destiné uniquement au marché français.

Hier zwei seltene Nachbildungen der Pariser Metro: ein Uhrwerkzug in Spur 0 von 1907 und ein Elektrozug in Spur 1 von 1906. Der Stromabnehmer des Triebwagens läuft auf einer äußeren Stromschiene – für ein Spielzeug sehr ungewöhnlich. Diese Nachbildungen waren wahrscheinlich nur für den französischen Markt bestimmt.

Probably introduced in 1905, this electric tram is for gauge 1. At the time of its introduction its factory price was comparable to the prices of Märklin's locomotives; it was not commercially successful, and is therefore rare nowadays. It was offered in clockwork as well as electric form, in both gauge 0 and gauge 1, and with different inscriptions for the export markets.

Probablement introduit en 1905, ce tramway électrique est en écartement 1. A l'époque de son lancement son prix à la sortie d'usine était comparable à celui des locomotives; il n'obtint cependant pas de succès commercial et il est donc rare aujourd'hui. Produit en versions mécanique et électrique, et dans les deux écartement 0 et 1, il avait des inscriptions différentes suivant les pays.

Diese elektrische Straßenbahn in Spur 1 wurde wahrscheinlich 1905 eingeführt. Der Preis ab Werk entsprach denen der Märklin-Lokomotiven. Sie war kein kommerzieller Erfolg und ist daher heute recht selten. Sie wurde als Elektro- und Uhrwerkausführung in den Spuren 0 und 1 und mit fremdsprachigen Beschriftungen für den Export angeboten.

TRAMWAY ÉLÉCTRIQUE

Spain was the destination of this gauge 1 clockwork tram of about 1904. It was part of a boxed set comprising two tram cars, two circuits and an interesting stopping system.

C'est à l'Espagne qu'était destiné ce tramway mécanique de 1904 en écartement 1. Il faisait partie d'un coffret comprenant deux tramways, deux circuits parallèles et un curieux système d'arrêt.

Diese um 1904 hergestellte Uhrwerk-Straßenbahn in Spur 1 war für Spanien bestimmt. Sie gehörte zu einem Satz mit zwei Straßenbahnwagen, zwei Strecken und einer ungewöhnlichen Bremsanlage.

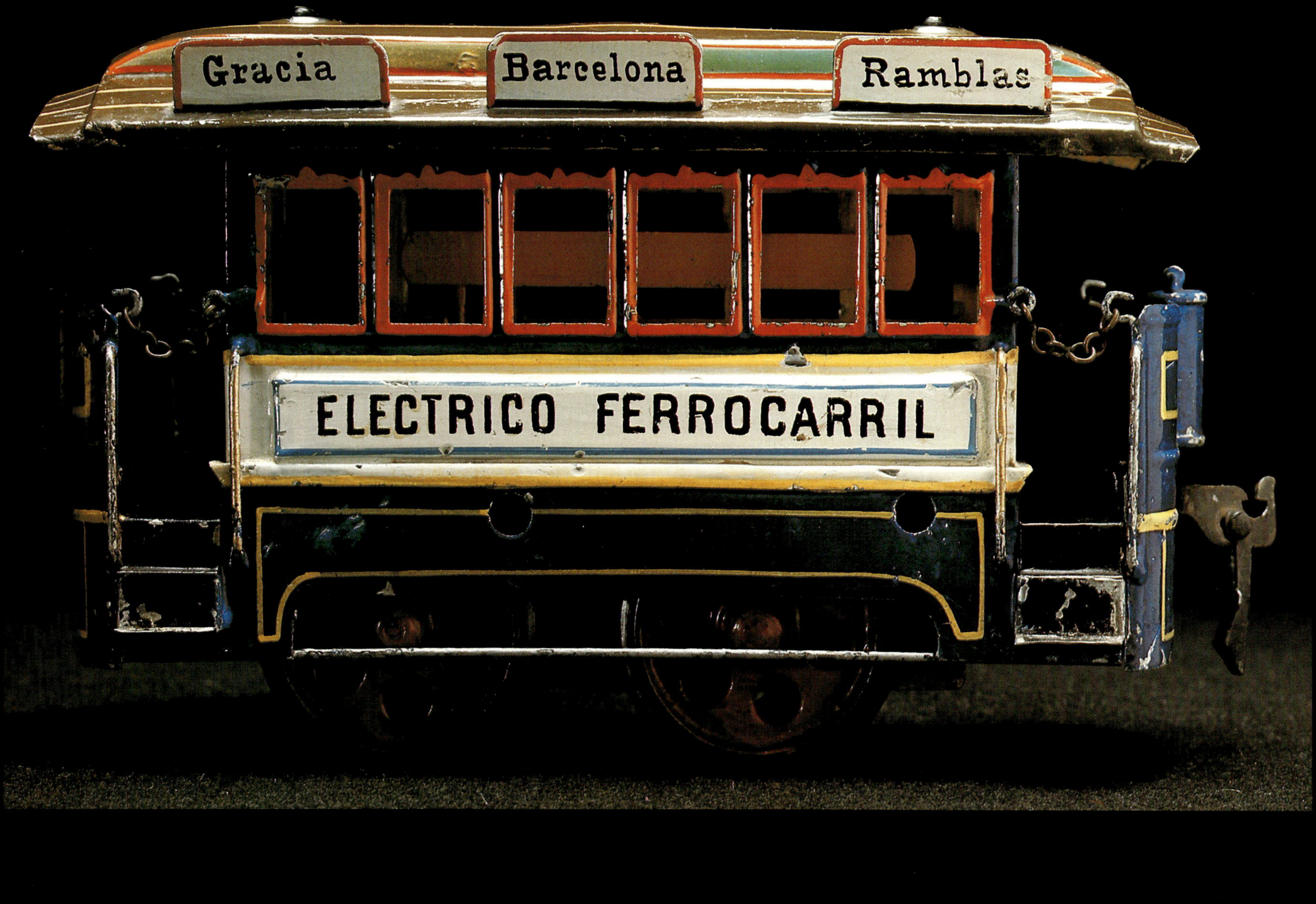

This gauge 1 toy was issued between 1904 and 1906 with a clockwork motor, despite being called an 'electric' tramway. Its destination boards are unmarked.
A rarer version exists in gauge 0.

Malgré l'inscription 'Electric Tramway' ce modèle en écartement 1 fut produit entre 1904 et 1906 avec un moteur mécanique. Ses panneaux ne portent aucune indication de destination. Une version plus rare existe en écartement 0.

Dieses zwischen 1904 und 1906 verkaufte Spielzeug in Spur 1 wurde zwar als elektrische Straßenbahn bezeichnet, besaß aber ein Federwerk. Die Fahrzieltafeln waren unbeschriftet. Eine seltenere Ausführung hat die Spur 0.

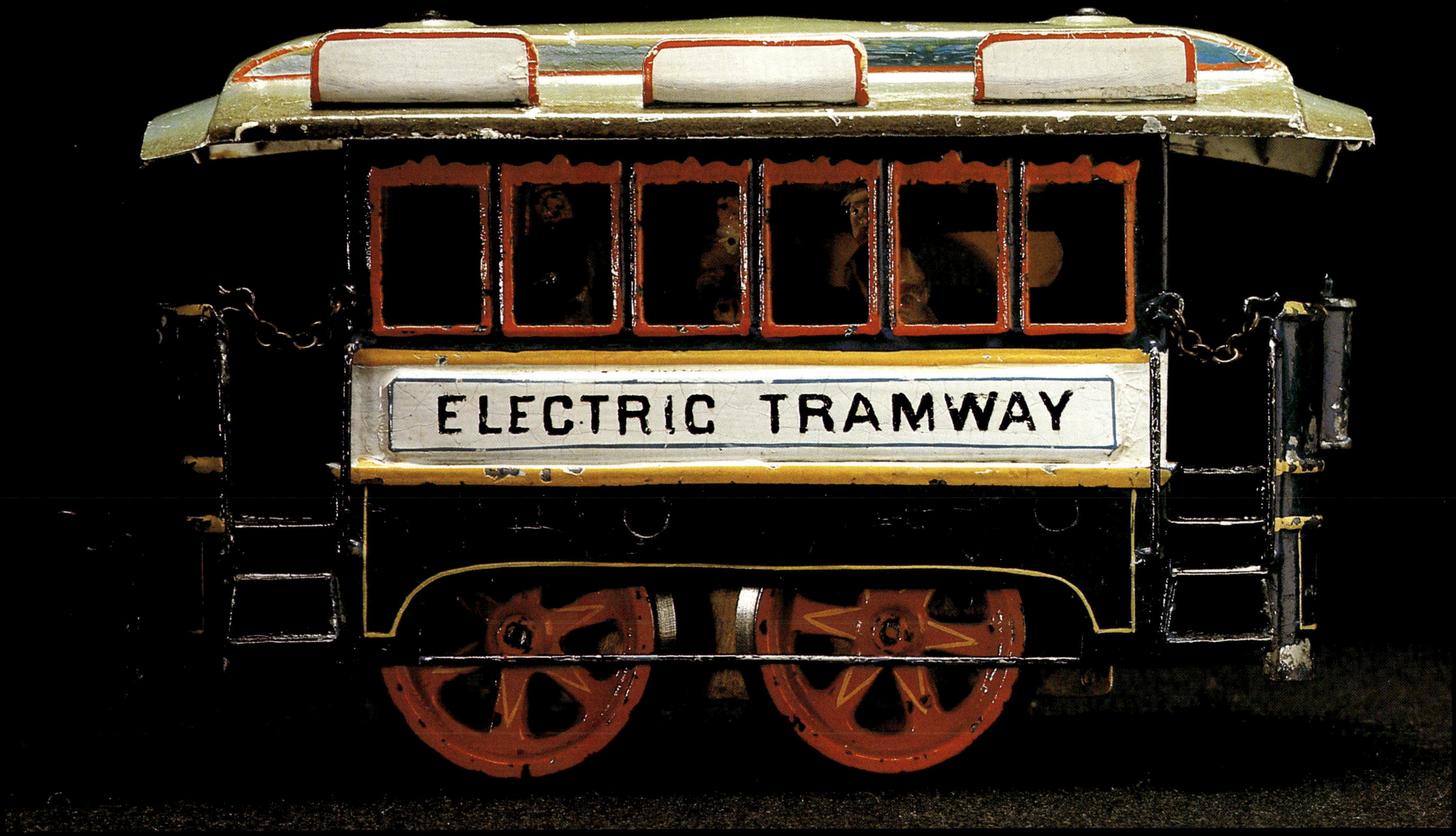

This rare gauge 1 trailer tram-car of 1909 was a prototype for America; no example with a motor has ever been found. Unusual features include the doors which open inwards.

Ce rare tramway en écartement 1 de 1909 était un prototype pour les Etats-Unis; aucun exemplaire connu ne possède de moteur. Parmi d'autres curieuses caractéristiques les portes sur ce modèle s'ouvrent de l'intérieur.

Dieser seltene Straßenbahnanhänger in Spur 1 von 1909 war ein Prototyp für Amerika. Sein motorisiertes Gegenstück ist unbekannt. Ungewöhnlich sind die nach innen zu öffnenden Türen.

On this 1905 gauge 1 tram the electric current was received not by the usual central rail but via a central roof rod – as it was on this toy's real-life counterparts.

Ce tramway de 1905 en écartement 1 était alimenté non par le rail central comme de coutume, mais par une perche au milieu de toit, à l'instar des véritables tramways.

Bei dieser Straßenbahn in Spur 1 von 1905 wurde der Fahrstrom nicht von einer Mittelschiene, sondern wie bei den echten Vorbildern über eine Stromabnehmerstange auf dem Dach abgenommen.

A key inserted in a hole in the centre of the roof activates this fine 1905 clockwork tram. Made in continental outline, this version in gauge 0 is much rarer that its counterpart in gauge 1. Thick tinplate and paint have been used in the construction and decoration of a relatively small toy destined for the English market.

Une clef inserée dans un trou au centre du toit fait fonctionner ce beau tramway mécanique de 1905. De ligne continentale, cette version en écartement 0 est plus rare que celle en écartement 1. Un épais fer-blanc et une peinture pareillement épaisse ont été utilisés pour la construction et la décoration de ce jouet relativement petit qui était destiné au marché britannique.

Diese hübsche Straßenbahn von 1905 wurde mit einem in das Dach eingesteckten Schlüssel aufgezogen. Diese nach kontinentalen Vorbildern in Spur 0 gebaute Ausführung ist wesentlich seltener als das Gegenstück in Spur 1. Für die Herstellung dieses relativ kleinen für den englischen Markt bestimmten Spielzeugs wurde starkes Blech benutzt.

Electric Tramway

This clockwork armoured train was featured in Märklin's catalogue of 1902 and in Gamage's catalogue of the same date. It was inspired by the English trains which transported gold during the Boer War in South Africa. This example is in gauge 1; the toy was also available in gauge 0 and gauge 2. Only clockwork models were offered. The two details on the left illustrate the train's interesting gun-firing mechanism; three hammers detonate a firing cap as the wagons

make contact with special points placed on the rails. Six separate explosions occur during one circuit of the train. The version produced for sale in America was supplied with a cow-catcher.

Ce train blindé mécanique figurait dans les catalogues Märklin et Gamage de 1902. Il fut inspiré par les trains anglais qui transportaient de l'or en Afrique du Sud pendant la guerre des Boers. Le modèle illustré ici est en écartement 1; il était aussi disponible en écartement 0 et 2; il n'existe qu'en version mécanique. A gauche, deux détails illustrent l'intéressant mécanisme de tir; trois marteaux font détonner des amorces au moment où les wagons entrent en contact avec des ergots qui étaient placés sur les rails. Il y a six détonations séparées par circuit complet. Le modèle américain était vendu avec le chasse-boeufs habituel.

Dieser Uhrwerk-Panzerzug erschien in den Katalogen von Märklin und Gamage's von 1902. Sein Vorbild waren die englischen Züge, die während des Burenkrieges in Südafrika Gold beförderten. Dieses Modell hat die Spurweite 1. Der Zug wurde aber auch in Spur 0 und 2 angeboten. Alle Ausführungen besaßen Federwerke. Die beiden Ausschnitte links zeigen den Schießmechanismus des Zuges: drei Hämmer treffen ein Zündhütchen, wenn die Wagen auf den Schienen angebrachte Kontakte berühren. Während einer Rundfahrt des Zuges wurde sechsmal geschossen. Die Ausführung für Amerika besaß einen Kuhfänger.

The clockwork pinchbar, or power car, was attached to the back of a train to boost its power. The 1908 French version and the 1907 German version are shown here; the English model in maroon or white and brown was inscribed 'power car'. Only Märklin manufactured this type of booster; gauge 0 and gauge 1 models were offered between 1907 and 1914.

Cette voiture motrice mécanique était attachée à l'arrière d'un train pour accroître sa puissance. Versions française de 1908 et allemande de 1907 sont illustrées ici; le modèle anglais aux couleurs bordeaux ou blanc et marron portait l'inscription 'Power Car'. Seul Märklin a fabriqué ce type de wagon-pousseur qui fut disponible en écartement 0 et 1 entre 1907 et 1914.

Der zusätzliche Federwerkantrieb wurde hinten an den Zug angekuppelt. Hier werden die deutsche Ausführung von 1907 und die französische von 1908 gezeigt. Das entweder kastanienbraune oder weißbraune englische Modell trug die Aufschrift „power car". Ein derartiger Zusatzantrieb wurde nur von Märklin hergestellt. Von 1907 bis 1914 wurden Modelle in den Spuren 0 und 1 angeboten.

Introduced in about 1908, this fine gauge 1 'beer car' graphically represents the transportation of the famous American beer 'Budweiser'. Märklin introduced other similar advertising wagons specifically for the American market at this time, including those for 'Tomato Ketchup', 'Schlitz' and 'Pabst Blue Ribbon'. The bold lettering on the early versions was hand-painted; on later examples it was lithographed.

Introduit vers 1908, ce très beau wagon représente un transport de la célèbre bière américaine 'Budwieser'. A la même époque Märklin lança sur le marché américain d'autres wagons publicitaires similaires pour les marques suivantes: 'Tomato Ketchup', 'Schlitz' et 'Pabst Blue Ribbon'. Sur les premières versions l'inscription était peinte à la main; par la suite elle fut lithographiée.

Dieser um 1908 eingeführte hübsche „Bierwagen" in Spur 1 zeigt, wie das berühmte amerikanische Budweiser Bier befördert wurde. Märklin brachte damals speziell für Amerika auch andere Waggons mit Werbeaufschriften heraus, so z.B. „Tomato Ketchup", „Schlitz" und „Pabst Blue Ribbon". Die ursprünglichen fetten Handbeschriftungen wurden später durch Druck ersetzt.

USCH
ISER
TLED BEERS.
3600
STLR.
G.CO.
3600
CAPY. 60000
ANHEUSER - BUSCH
BEER CAR.

These versions of the Red Cross carriage were offered from 1902 to 1912. The example shown below was made in 1908, that on the right in 1907. Gauges 0, 1 and 2 models were available. The interiors included stretchers, opening medicine chests, wounded figures and a stove with a chimney. The smaller carriage's windows are similar to those of some Märklin stations.

Ces versions du wagon Croix-Rouge furent mises sur le marché entre 1902 et 1912. Le modèle ci-dessous fut fabriqué en 1908, celui à droite en 1907. Ce wagon existait en écartement 0, 1 et 2. L'intérieur comprenait brancards, placard à pharmacie, blessés et un poêle avec cheminée. Les fenêtres du wagon, bien que plus petites, sont semblables à celles de certaines gares Märklin.

Diese Rotkreuzwagen wurden von 1902 bis 1912 angeboten. Die untere Version stammt von 1908, die rechte von 1907. Es wurden Modelle in den Spurweiten 0, 1 und 2 vertrieben. Zur Innenausstattung gehörten Bahren, Verbandschränke, Verwundete und ein Ofen. Die Fenster des kleineren Wagens sind die gleichen wie bei einigen Märklin-Bahnhöfen.

Snow ploughs were made between 1902 and 1912. The gauge 0 and gauge 1 versions are shown here; the gauge 2 model is rarer. Each plough was sold in a set which included figures and the large lantern; this is often missing nowadays.

Les chasses-neiges furent fabriqués de 1902 à 1912. Les versions en écartement 0 et 1 sont illustrées ici; la version en écartement 2 est la plus rare. Chaque wagon chasse-neige était vendu complet avec personnages et lanterne, qui souvent aujourd'hui sont manquants. Ce wagon était attaché à l'avant de la locomotive afin de déblayer la voie.

Schneepflüge – hier die Modelle in Spur 0 und 1 – wurden von 1902 bis 1912 hergestellt. Die Spur 2 ist seltener. Jeder Pflug wurde als Satz mit Figuren und einer großen Laterne verkauft, die heute meist fehlt.

The snow plough was attached to the front of a locomotive to clear the track.

Der Schneepflug wurde vor der Lok zum Räumen der Strecke angekuppelt.

St. E.V.
21.
St. E.V
20

This gauge 1 heating wagon of 1906 is one of the rarest and most original of Märklin's carriages; it was made in gauge 1 only. It contains a working boiler with a spirit burner (shown in action in the detail) and a steam outlet which heats the adjacent carriages. It is a classic example of Märklin's commitment to detail.

Ce wagon chaufferie de 1906 en écartement 1 est parmi les plus rares et les plus originaux des wagons Märklin; il n'existe qu'en écartement 1. Il contient une véritable chaudière marchant avec un brûleur à alcool (que l'on voit en fonctionnement sur le détail) et une sortie de vapeur pour chauffer les voitures voisines. Voici un exemple typique de la finition Märklin.

Dieser Heizungswagen in Spur 1 von 1906 ist einer der seltensten und originellsten Wagen von Märklin und wurde nur in dieser Spur hergestellt. Er enthält einen funktionsfähigen Dampfkessel mit einer Spiritusflamme (im Ausschnitt im Betrieb dargestellt) und eine Dampfleitung zur Beheizung der übrigen Wagen. Ein klassisches Beispiel der Detailtreue Märklins!

These horse transporter wagons in gauge 0 and gauge 1 were made for the English market between 1907 and 1919. The livery is that of the GNR (Great Northern Railway). Features of the original trucks are accurately reproduced, particularly the special doors which allow the horses, also supplied by Märklin, to enter and leave with ease on the solidly constructed ramp.

Ces transports de chevaux en écartement 0 et 1 furent fabriqués pour le marché britannique entre 1907 et 1919. Les couleurs sont celles de la GNR (Great Northern Railway). Les traits caratéristiques des wagons originaux sont reproduits avec exactitude, telles les portes spéciales qui permettent aux chevaux, aussi fournis par Märklin, d'entrer et sortir avec facilité en empruntant la solide rampe d'accès.

Diese Pferdetransporter in Spur 0 und 1 wurden zwischen 1907 und 1919 für den englischen Markt hergestellt. Die Farben sind die der GNR (Great Northern Railway). Die Merkmale der Originalvorlagen sind genau wiedergegeben, so besonders die Spezialtüren, durch die die ebenfalls von Märklin stammenden Pferde über die robusten Rampen leicht ein- und ausgeladen werden konnten.

The interior of this interesting but sinister gauge 1 prison wagon includes fittings such as individual cells and chairs for the guards, as can be seen below. The example shown here was designed for sale in Germany; other versions, with their inscriptions altered for the different export markets, were available in three gauges. The toy was made from 1907 to 1914.

L'intérieur de cet intéressant mais sinistre wagon pénitencier en écartement 1 comprend des aménagements telles cellules individuelles et chaises pour les gardiens, comme on peut l'observer ci-dessous. Le modèle illustré ici était destiné au marché allemand; d'autres versions, avec leurs inscriptions modifiées suivant les pays d'exportation, étaient disponibles en trois écartements. Ce jouet fut fabriqué de 1907 à 1914.

Zur Innenausstattung dieses interessanten, doch düsteren Gefangenenwagens in Spur 1 gehören Einzelzellen sowie Stühle für die Wärter. Das hier gezeigte Modell war für Deutschland bestimmt. Die in drei Spurweiten hergestellten Exportmodelle besaßen Beschriftungen je nach dem Bestimmungsland. Die Wagen wurden von 1907 bis 1914 gebaut.

GEFANGENEN WAGEN

Though there are no interior fittings in this gauge 1 ventilated wagon made for the German market in about 1909, the originals were probably used to transport fruit and vegetables.

Bien que ce wagon avec volets d'aération en écartement 1, fabriqué pour le marché allemand, ne comprenne pas d'aménagement intérieur, il est fort probable que l'original ait été destiné au transport de fruits et légumes.

Dieser um 1909 für den deutschen Markt hergestellte belüftete Waggon in Spur 1 besitzt keine Innenausstattung. Die echten Vorbilder wurden wahrscheinlich zur Beförderung von Obst und Gemüse verwendet.

A different ventilated wagon, based on the fruit and vegetable truck, was made for England between 1907 and 1914. Midland Railway and LNWR versions, including this gauge 0 example, were available.

Cette version différente du wagon aéré, basée sur le transport de fruits et légumes, fut fabriquée pour l'Angleterre entre 1907 et 1914. Des versions Midland Railway et LNWR étaient aussi disponibles.

Zwischen 1907 und 1914 wurde eine Ausführung für England gebaut, für die der Obst-waggon das Vorbild war. Midland Railway-und LNWR-Ausführungen einschließlich dieses Modells in Spur 0 wurden angeboten.

'Kaiser' saloon cars in three gauges made from 1906 to 1914 are shown on the right. The gauge 2 version was offered in 1906, the gauge 1 from 1908 to about 1914, and the rarest, gauge 0, in 1907. The imperial crowns on the coaches' bodywork suggest that these cream and blue carriages were part of a 'Kaiserzug', a royal train. The bright blue version shown above has the same outline but is without the decorative crest. It was offered for sale in France.

A droite les trois versions du wagon-salon 'Kaiser' en écartements 0, 1 et 2. La version en écartement 2 fut offerte en 1906, celle en 1 de 1908 aux alentours de 1914, et la plus rare, celle en 0, en 1907. Les couronnes impériales sur les carosseries des wagons suggèrent que ces voitures crème et bleu faisaient partie d'un train royal ('Kaiserzug'). La version en bleu ci-dessus possède la même ligne mais mais ne porte pas d'armoiries.

Rechts in drei Spurweiten zwischen 1906 und 1914 gebaute Salonwagen. Das Modell in Spur 2 stammt von 1906, die Spur 1 von 1908 bis etwa 1914 und die seltene Spur 0 von 1907. Die Kaiserkronen auf den Seiten deuten an, daß die cremefarbenen und blauen Wagen zu einem Kaiserzug gehörten. Das für Frankreich bestimmte obige hellblaue Modell hat den gleichen Aufbau, jedoch ohne das kaiserliche Wappen.

The charm of this ornate gauge 1 rack rail carriage stems from the quality of design and production, both characteristic of Märklin's early work. Made between 1904 and 1912 in three gauges, it shows various contemporary manufacturing techniques in use. From the pierced railing protecting the open corridor to the embossed columns supporting the roof, it is a delightful toy.

Le charme de ce wagon ouvragé en écartement 1 pour train à crémaillère provient de la qualité du dessin et de la fabrication, tous deux caractéristiques de la production Märklin des premières années. Fabriqué de 1904 à 1912 en trois écartements, il est exemplaire de l'utilisation des techniques de l'époque. De la ballustrade ajourée protégeant le couloir promenoir aux colonnes estampées supportant le toit, c'est un jouet charmant.

Der Charme dieses schmuckvollen Zahnradbahnwagens in Spur 1 ist die für Märklins frühe Schöpfungen typische Qualität von Entwurf und Ausführung. Die zwischen 1904 und 1912 in drei Spuren gebauten Wagen demonstrieren die zeitgenössischen Fertigungsverfahren. Ein hübsches Spielzeug – vom durchbrochenen Schutzgeländer am Gang bis zu den geprägten Dachsäulen.

This 1907 double-decker carriage exists in both gauge 0 and gauge 1 form; there is no known example in gauge 2. The upper level is rarely found; many collectors are obliged to reproduce it, pending the discovery of an original. Made specifically for France, this toy was based on carriages that were used on the suburban railway lines of Paris.

Ce wagon à impériale de 1907 existe en écartements 0 et 1; il n'y a pas d'exemplaire connu en écartement 2. L'impériale est difficile à trouver: certains collectionneurs sont obligés de la reproduire eux-mêmes en attendant la découverte d'un original. Fabriqué spécialement pour la France, ce jouet trouve son modèle dans des voitures utilisées sur les lignes de la banlieue parisienne.

Dieser zweistöckige Personenwagen von 1907 existiert in den Spuren 0 und 1, während eine Spur 2 nicht bekannt ist. Das Oberdeck ist sehr selten, und die meisten Sammler müssen sich mit Reproduktionen begnügen, solange sie kein Original entdecken. Die Vorbilder dieser speziell für Frankreich gebauten Wagen liefen auf Pariser Vorortstrecken.

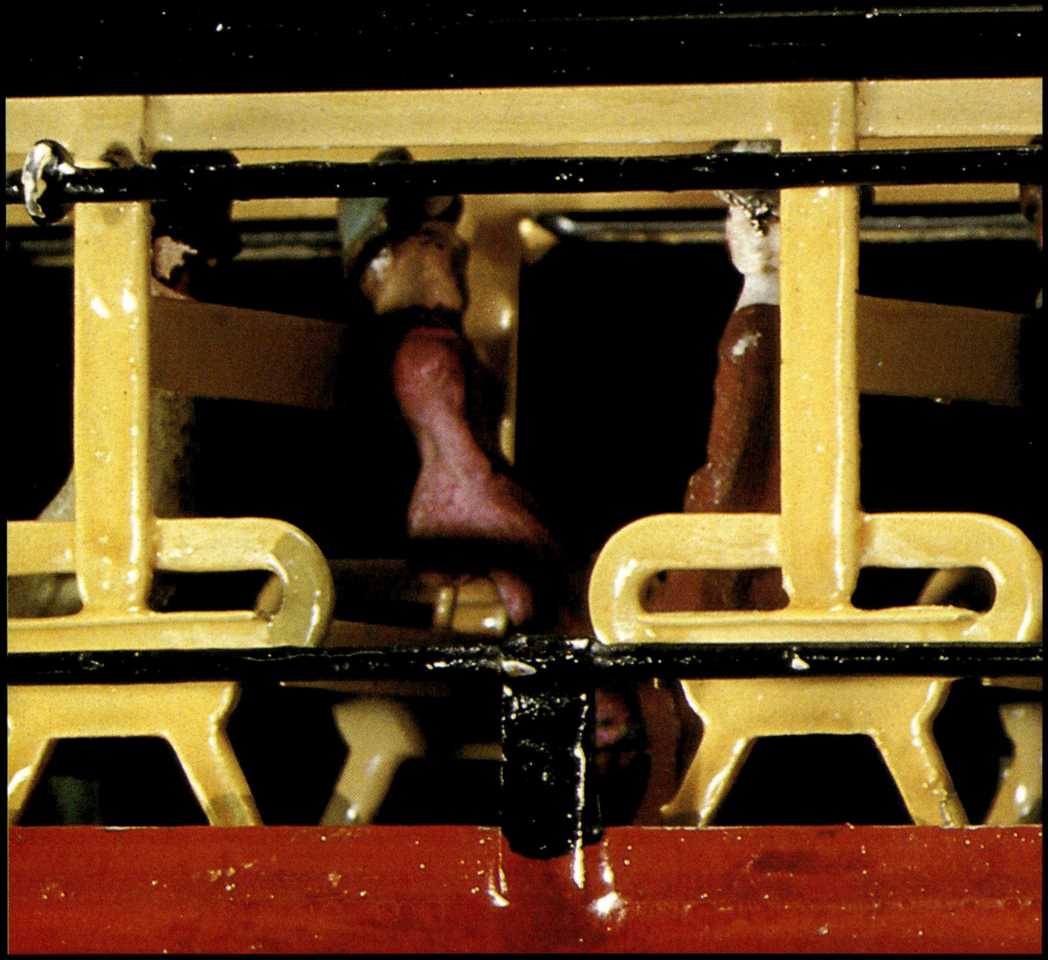

A clockwork inspection car is shown here with its two rare original figures; the foreman sat in the front inspecting the track. The movement gives the impression that the driver controls the progress of the car along the track. This gauge 0 example was made in about 1907. The model was available in three gauges; gauge 2 is the rarest.

Une draisine mécanique est ilustrée ici avec ses deux rares personnages d'origine; le contremaître s'asseyait à l'avant afin d'inspecter les rails. Le mouvement donne l'impression que le conducteur contrôle la marche de la draisine le long des rails. Cet exemplaire en écartement 0 fut fabriqué vers 1907. Le modèle était disponible en trois écartements; l'écartement 2 est le plus rare.

Die Abbildung zeigt eine Uhrwerk-Draisine mit ihren zwei seltenen Figuren. Der Streckenkontrolleur sitzt vorne. Die Draisine bewegt sich so, als ob der Fahrer die Vorwärtsbewegung bewirkt. Dieses Modell in Spur 0 wurde um 1907 gebaut. Von den drei Spuren ist die Spur 2 am seltensten.

Three figures (two railway workers and a uniformed foreman) were supplied with this 1907 gauge 1 example of the clockwork inspection car. The large wheels were made specifically for this toy. Finely engraved patterns cover the glass roof of the platform; these were probably inspired by the work of Baltard or the Art Nouveau style.

Deux ouvriers et un contremaître étaient fournis avec cet exemplaire de 1907 de la draisine mécanique en écartement 1. Les grandes roues ont été spécialement fabriquées pour ce modèle. Des motifs finement gravés ornent le toit de verre du quai; ils furent probablement inspirés ou par l'oeuvre de Baltard ou par le style Art Nouveau.

Zu dieser Uhrwerk-Draisine von 1907 in Spur 1 gehörten drei Figuren (zwei Streckenarbeiter und ein uniformierter Vorarbeiter). Die großen Räder wurden speziell für dieses Spielzeug geschaffen. Die feinen Gravuren über dem Glasdach des Bahnsteigs gehen wahrscheinlich auf die Arbeiten Baltards oder den Art Nouveau-Stil zurück.

This gauge 0 removals van was made for export in three gauges from 1903 to 1910. Two horses pulled it to the station where it was placed on a rail transporter.

Cette roulotte de déménagement fut fabriquée pour l'exportation de 1903 à 1910 en trois écartements. Deux chevaux l'entraînaient à la gare ou elle était alors placée sur un wagon de transport.

Dieser Möbelwagen in Spurweite 0 wurde von 1903 bis 1910 für den Export in drei Spuren angefertigt. Zwei Pferde zogen ihn zum Bahnhof und dort auf einen Flachwagen.

Circuses often used rail transport in the 1900s. This attractive gauge 0 caravan, offered between 1903 and 1910, has fine interior fittings. The inscription is in traditional circus lettering.

Dans les années 1900 les cirques ambulants utilisaient souvent le transport par rail. Cette attrayante roulotte en écartement 0 qui fut offerte entre 1903 et 1910 possède de beaux aménagements intérieurs.

Zirkusse benutzten um 1900 oft die Bahn. Dieser von 1903 bis 1910 gebaute Zirkuswagen in Spur 0 hat eine hübsche Innengestaltung. Die Beschriftung geht auf traditionelle Zirkusbezeichnungen zurück.

The Army Corps mobile kitchen was made in blue from 1904 to 1914; from 1919 it was produced in pale green. It was marketed in several countries and was probably available either with the transporter or separately. The Märklin figure shown here was not supplied with the toy; a figure in military uniform would probably have accompanied it originally.

La cuisine mobile 'Armée Corps' fut produite en bleu de 1904 à 1914; à partir de 1919 elle le fut en vert pâle. Elle fut exportée dans plusieurs pays et était probablement disponible avec ou sans wagon de transport. Le personnage Märklin que l'on voit ici n'était pas fourni avec le jouet; un personnage militaire accompagnait probablement l'original.

Die Feldküche des Armee Corps war von 1904 bis 1914 blau und ab 1919 blaßgrün. Sie wurde in mehreren Ländern verkauft und wahrscheinlich mit und ohne den Güterwagen angeboten. Die hier gezeigte Märklin-Figur wurde nicht mitgeliefert. Wahrscheinlich gehörte ursprünglich ein Soldat dazu.

I. Armee - Corps.

An unusually complete group of gauge 0 pieces is shown here crossing a finely produced Märklin bridge of about 1908. The lattice-work bridge is a good example of Märklin's preoccupation with environmental and architectural surroundings of the period. It is possibly based on a contemporary bridge designed by the famous engineer Eiffel, who specialized in bridges with lattice constructions.

Une rame exceptionnellement complète, en écartement 0, passe sur un beau pont Märklin de 1908. Ce pont en treillis est un bon exemple de la préoccupation de Märklin avec l'environnement et l'architecture de l'époque. Il est possible qu'il ait été inspiré par les ponts en treillis de Jean Eiffel.

Ein ungewöhnlich kompletter Zug in Spur 0 überquert hier eine sorgfältig gestaltete Märklin-Brücke von 1908. Die Gitterträgerbrücke beweist Märklins Detailtreue zu den architektonischen Vorbildern seiner Umwelt. Vorbild war wohl eine Brücke des berühmten Konstrukteurs Eiffel, dessen Spezialität Gitterträgerbrücken waren.

These standard lamps, available between 1904 and 1910, are just two of Märklin's wide range of lighting accessories. The double lamp is run by electricity, the single lamp by oil.

Ces lampadaires, disponibles entre 1904 et 1910, sont deux exemples parmi la vaste gamme d'accessoires d'éclairage de Märklin. Le double marche à l'électricité, le simple à l'huile.

Diese zwischen 1904 und 1910 angebotenen Bogenlampen sind nur zwei von den vielen Märklin-Leuchtkörpern. Die Doppellaterne ist eine elektrische, die Einzellaterne eine Öllampe.

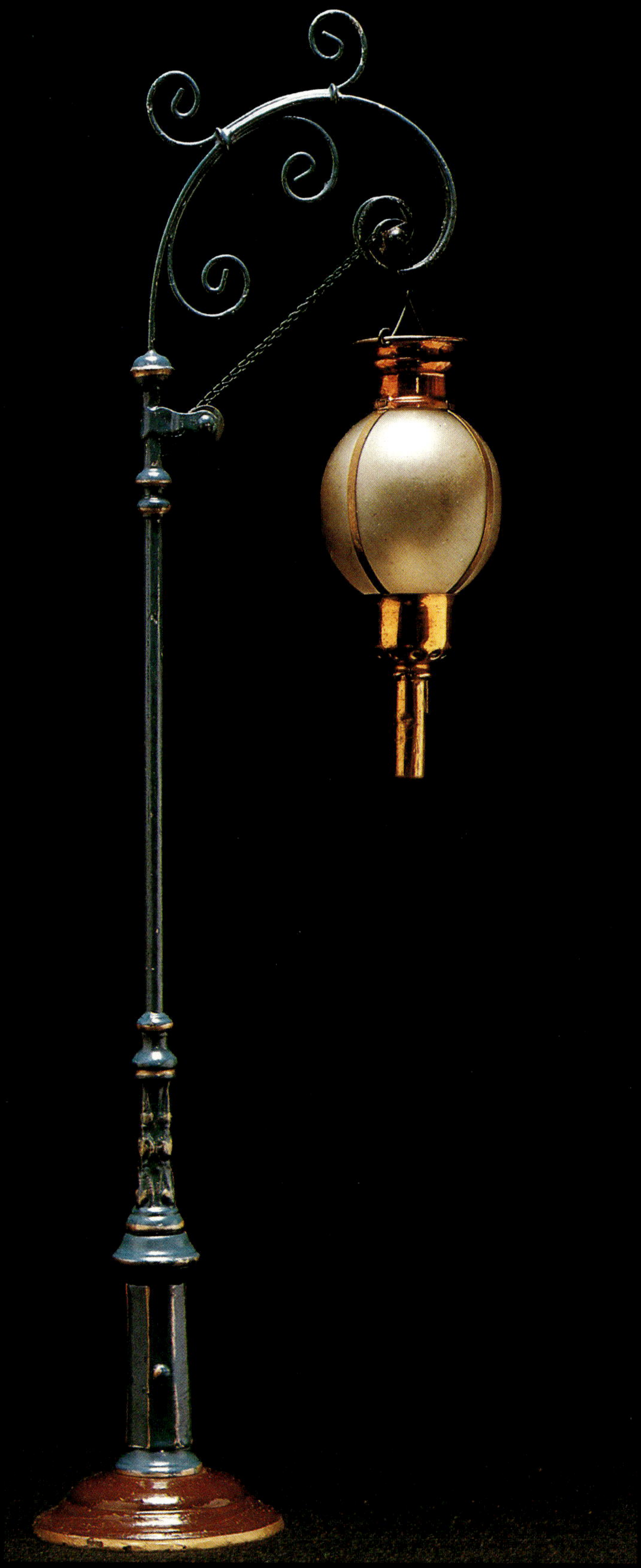

These indicator boards were made between 1904 and 1910 for England. The blue example has a working bell and oil lamp; the green board was probably for an early London underground line.

Ces panneaux indicateurs furent fabriqués entre 1904 et 1910 pour l'Angleterre. L'exemplaire bleu a une cloche et une lampe à huile; le vert était probablement destiné à l'un des premiers métros londonniens.

Diese Anzeigetafeln wurden zwischen 1904 und 1910 für England hergestellt. Die blaue hat eine funktionsfähige Glocke und eine Öllampe, und die grüne war wahrscheinlich für die erste U-Bahn in London bestimmt.

Märklin's versatility and commitment to detail are shown once more in this selection of accessories. Illustrated from left to right are a cattle weighing scale of about 1905, a porter's trolley, laden with some superbly detailed luggage, made in 1902 and the ubiquitous French gentleman's *pissoir* of 1907.

Cette sélection d'accessoires illustre la variété de Märklin tout autant que l'attention apportée au détail. De gauche à droite: une bascule à bétail datant de vers 1905, un charriot de porteur fabriqué en 1902 et rempli de bagages remarquablement finis, et une vespasienne datant de 1907.

Dieses Zubehör zeigt erneut Märklins Vielseitigkeit und Detailtreue. Von links nach rechts: eine Viehwaage von etwa 1905, ein Gepäckkarren von 1902 mit sorgfältig ausgeführten Gepäckstücken und ein typisches französisches *pissoir* von 1907.

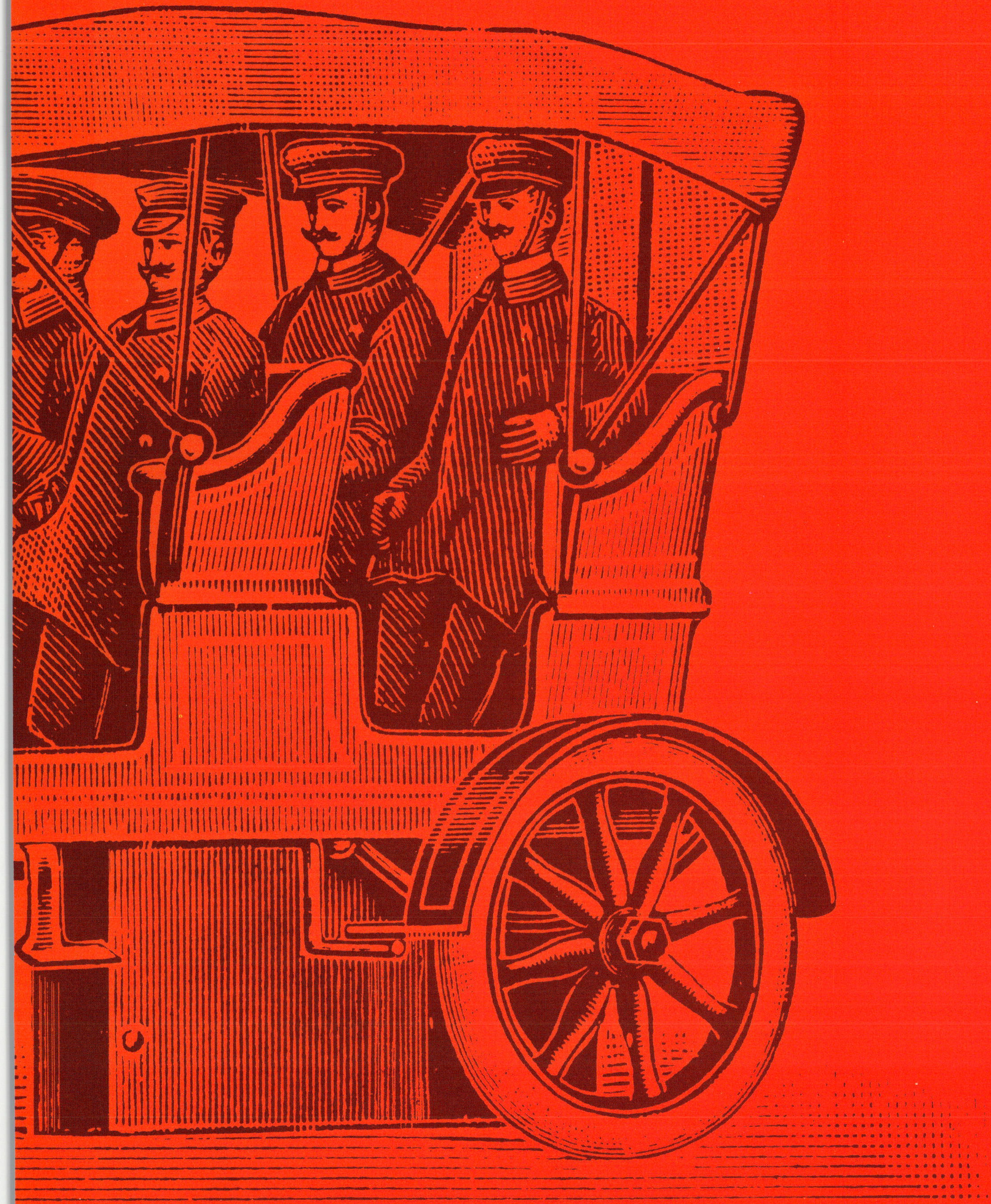

AUTOMOBILES
AUTOS

Compared with the other ranges Märklin's automobiles are nowadays extremely rare. Before 1914 three basic sizes of limousine were produced in several colours.
This medium-sized light blue clockwork limousine was offered between about 1906 and 1914; the luggage was sold separately. The brakes, which operate on the back wheels, are activated by a side lever. The fine original lamps, polished brass and nickel-plated, can be seen in the detail; such lamps are a common feature of Märklin cars produced between 1904 and 1919.

Märklin-Automobile sind heute im Vergleich zu den anderen Erzeugnissen äußerst selten. Vor 1914 wurden im wesentlichen drei Größen in mehreren Farbabwandlungen gebaut. Diese mittelgroße Aufziehlimousine wurde zwischen ca. 1906 und 1914 angeboten. Das Gepäck wurde separat verkauft. Die Hinterradbremsen wurden durch einen seitlichen Hebel betätigt. Der Ausschnitt zeigt die für die Märklin-Autos zwischen 1904 und 1919 typischen sorgfältig gefertigten Scheinwerfer aus poliertem Messing bzw. vernickeltem Blech.

Comparées aux autres gammes, les automobiles Märklin sont aujourd'hui très difficiles à trouver. Avant 1914 trois tailles de limousine étaient produites en différentes couleurs.
Cette limousine bleu-clair de taille moyenne fut vendue entre les environs de 1906 et 1914; les bagages étaient vendus à part. Le frein, qui agit sur la roue arrière est actionné par un levier latéral. Le détail illustre les lanternes d'origine, en laiton poli et nickelé. De telles lanternes équipent presque toutes les voitures Märklin de 1904 à 1919.

Apart from a different colour scheme, this clockwork automobile of 1906 is almost identical to that shown on the previous pages. The addition of the number '5' on the side of the body probably indicates that it was designed as a taxi. One of the lamps, the steering column and the brake lever can be seen clearly on the left.

A part la couleur différente, cette automobile mécanique de 1906 est quasiment identique au modèle précédent. Le numéro '5' ajouté sur la porte indique probablement que nous avons affaire ici à un taxi. On peut voir clairement à gauche l'une des lanternes, la colonne de direction et le levier de frein.

Abgesehen von der anderen Farbgebung ist dieses Aufzieh-Auto von 1906 fast identisch mit dem auf den vorigen Seiten. Die Zahl „5“ am Aufbau bedeutet, daß es wahrscheinlich ursprünglich als Taxi gedacht war. Im Ausschnitt links sind deutlich ein Scheinwerfer, die Lenkradsäule und die Handbremse zu erkennen.

The largest size of the Märklin clockwork automobile range of 1908 is illustrated here. Exceptionally high standards of manufacture are evident in this saloon's cast-iron wheels, gilded luggage carrier, fine radiator and its spare tyre, attached to the roof with leather straps. A 1904 fountain and a rare battery-run street lamp complete the picture.

Illustrée ici, la plus grande des automobiles mécaniques de Märklin, datant de 1908. Cette berline témoigne d'une exceptionnelle qualité de fabrication ainsi qu'en font preuve les roues en fonte, le porte-bagage arrière doré, la superbe calandre de radiateur et le pneu de secours attaché par trois lanières de cuir sur le toit. Une fontaine de 1904 et un rare réverbère fonctionnant à pile complètent la scène.

Dies ist das größte von Märklins Aufzieh-Autos von 1908. Die außergewöhnlich hohe Qualität dieser Limousine beweisen die Gußeisenfelgen, der vergoldete Gepäckträger, die attraktive Kühlerkulisse und der mit einem Lederriemen auf dem Dach befestigte Reservereifen. Ein Brunnen von 1904 und eine seltene batteriebetriebene Laterne runden das Bild ab.

Petroleu

This clockwork automobile of about 1905 to 1910 is an example of the smallest of the three sizes of limousine made by Märklin before the First World War. It makes an interesting comparison with the two larger sizes; the medium size tends to be that most favoured by collectors. The sturdy steering wheel and column can be observed above.

Cette automobile mécanique (environs 1905-1910) est un exemple de la plus petite des trois tailles de limousines fabriquées par Märklin avant la Première Guerre Mondiale. On peut faire une comparaison intéressante avec les deux plus grandes tailles. La taille moyenne semble être celle préferée des collectionneurs. Peuvent être observés ci-dessus les robustes volant et colonne de direction.

Dieses Uhrwerk-Auto aus der Zeit von ca. 1905 bis 1910 ist die kleinste der drei von Märklin vor dem I. Weltkrieg hergestellten Limousinen. Interessant ist der Vergleich mit den beiden größeren Typen, von denen die mittlere bei den Sammlern am beliebtesten zu sein scheint. Man beachte die Stabilität von Lenkrad und -säule.

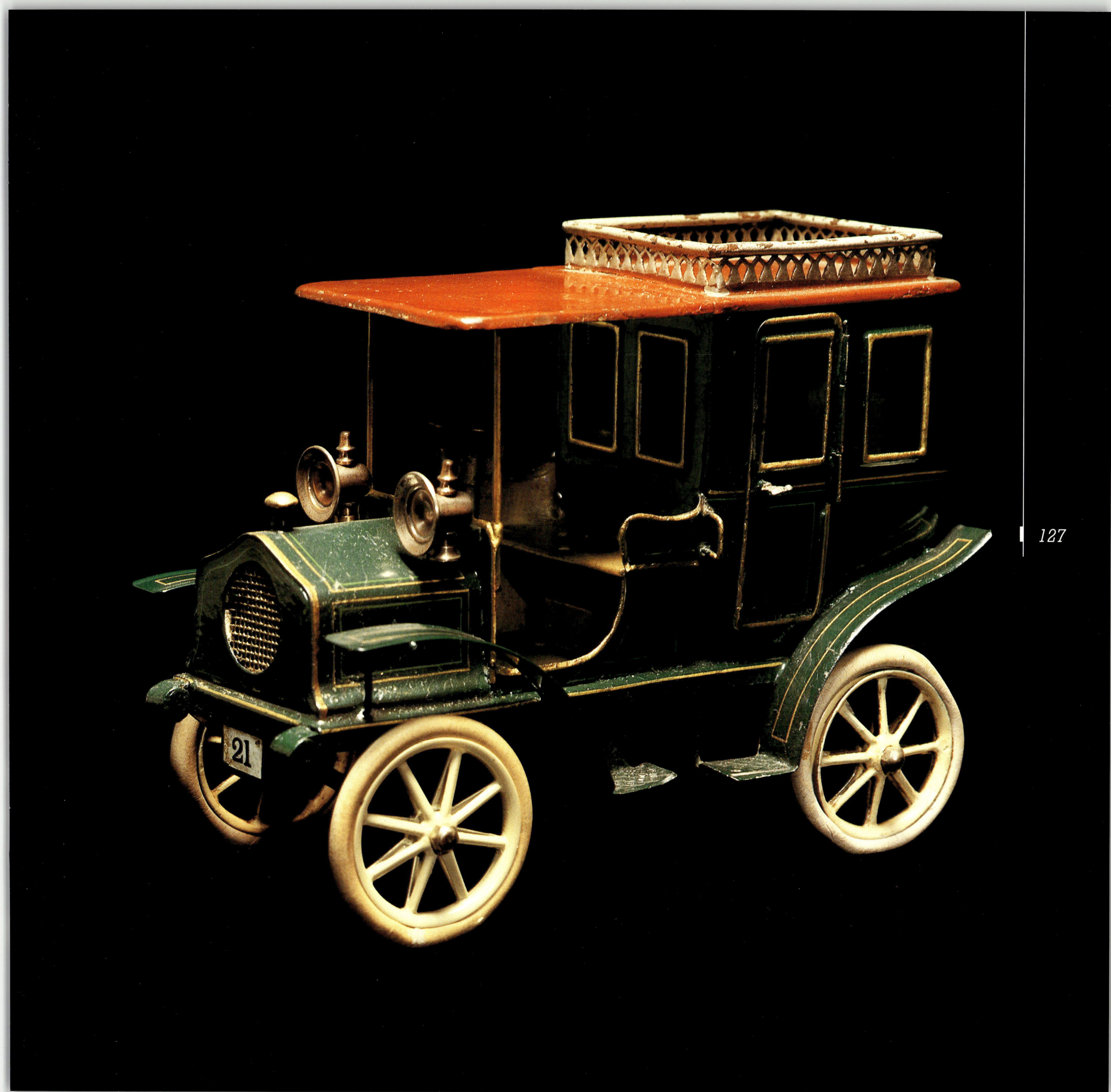

This handsome clockwork four seat open tourer first appeared in the Märklin catalogue of 1909. Four nickel-plated lamps, white rubber tyres, a complex geared steering system and embossed seats (imitating button upholstery)

Cette charmante voiture de tourisme mécanique à quatre places fit sa première apparition dans le catalogue Märklin de 1909. Quatre lanternes nickelées, des pneus de caoutchouc blanc, un système complexe de changement

Dieser hübsche viersitzige offene Tourenwagen mit Federwerk erschien zuerst im Märklin-Katalog von 1909. Dieser größte der drei Wagentypen hatte vier vernickelte Scheinwerfer, Weißwandreifen, eine recht komplizierte

are features of this car, the largest of the three sizes available. The bisque figures, two dressed in particularly appropriate motoring costumes, were probably made by Carette.

de vitesses et des sièges estampés (imitant le capitonnage à rembourage piqué) figurent sur cette voiture. Les deux figurines, habillées en costume d'automobiliste, furent probablement fabriquées par Carette.

Lenkung und imitierte Polstersitze. Die Figuren, zwei davon in zünftiger Autofahrerkleidung, stammen wahrscheinlich von Carette.

BOATS
BATEAUX
SCHIFFE

The rare clockwork paddle-boat 'Rotterdam' does not appear to have been illustrated in any Märklin catalogue. The boat's deck painting, however, is typical of the style used by Märklin around the turn of the century. The paddle-wheels, paddle-boxes, and mechanism appear in two other products offered in the 1904 catalogue (numbers 1069 and 1070). William Britain & Sons of England made the assortment of passengers shown below.

Il ne semble pas que ce rare bateau à aubes mécanique 'Rotterdam' ait été illustré dans les catalogues Märklin. La peinture du pont, cependant, est typique du style employé par Märklin au début du siècle. Les roues à aubes, tambour de roues et mécanisme figurent dans deux autres productions en vente dans le catalogue de 1904 (numéros 1069 et 1070). L'assortiment varié de passagers est fourni par la maison William Britain & Sons.

Dieser seltene Aufzieh-Raddampfer „Rotterdam" war anscheinend in keinem Märklin-Katalog enthalten. Die Decksbemalung ist jedoch typisch für Märklins Stil um die Jahrhundertwende. Die Schaufelräder, die Radkästen und der Mechanismus tauchen bei zwei anderen Erzeugnissen aus dem Katalog von 1904 auf (Nr. 1069 und 1070). Die im Ausschnitt gezeigten Passagiere stammen von William Britain & Sons aus England.

ROTTERDAM

An exciting clockwork toy of an Edwardian steam yacht, 'Jolanda' was made by Märklin for a twenty year period; it first appeared in the firm's catalogue of 1909 and was offered in a smaller size some time later. The fine quality of the paintwork of both the portholes and the cabin can be seen in the detail on the right.

Superbe jouet mécanique modelé sur un yacht anglais des années 1900, 'Jolanda' fut fabriqué par Marklin pendant une période de vingt ans; il apparut pour la première fois dans le catalogue de 1909 et fut par la suite proposé en une plus petite taille. Le détail à droite illustre la qualité de la peinture des hublots et des cabines.

Die Aufzieh-Dampfjacht „Jolanda" im edwardianischen Stil wurde von Märklin 20 Jahre lang gebaut und erschien zuerst im Katalog von 1909. Etwas später kam eine kleinere Variante heraus. Der Ausschnitt rechts läßt die sorgfältige Bemalung der Bullaugen und der Kabine erkennen.

'Diana' is among the most charming of the clockwork paddle-boats available from 1906. This toy epitomizes Märklin's high standards of characterization and production. The decorative paddle-wheel protectors, shown on the left, are detachable; consequently they are often missing.

'Diana' est parmi les plus charmants des bateaux à aubes mécaniques qui furent disponibles à partir de 1906. Ce jouet résume en fait le haut niveau de qualité et de production de Märklin. Les tambours de roue, à gauche, très décoratifs, sont détachables; en conséquence ils manquent souvent.

Die „Diana" ist einer der hübschesten Aufzieh-Raddampfer von 1906 und ist typisch für Märklins Qualitätsarbeit und Detailtreue. Die reich verzierten Radkästen (siehe Ausschnitt links) sind abnehmbar und gingen daher oft verloren.

DIANA

This clockwork 'Vedette', or pleasure boat, was made during the 1909-1914 period. It is one of the very few pieces illustrated in this book that has been extensively repainted; original processes and techniques were, however, used in the restoration. The shaping and embossing of the trawler-like hull would have been comparatively complicated. The passengers and the key are original.

Cette vedette de plaisance mécanique fut fabriquée pendant la période 1909-1914. C'est l'une des très rares pièces ilustrées dans ce livre qui ait été considérablement repeinte; elle fut toutefois restaurée selon les procédés et techniques de l'époque. Mettre en forme et estamper la coque, semblable à celle d'un chalutier, ne furent probablement pas choses simples. Passagers et clef sont d'origine.

Diese „Vedette", ein Ausflugsboot mit Federwerk, wurde von 1909 bis 1914 gebaut. Es ist eines der wenigen Objekte in diesem Buch, das weitgehend nachlackiert wurde, wenn auch mit den ursprünglichen Mitteln und Verfahren. Die Formung und Prägung des trawlerartigen Rumpfes waren sicherlich nicht einfach. Die Passagiere und der Schlüssel sind Originalstücke.

The steam-driven battleship 'Chicago' was first put on offer in the 1904 catalogue. The semi-flat sailors were made by the German firm of Heyde and were sold with many Märklin boats. The lead bases of the sailors could be slotted under small tabs attached to the deck; this ensured their stability when the boat was launched.

C'est dans le catalogue de 1904 qu'apparait pour la première fois ce navire de guerre à vapeur 'Chicago'. Les marins, quasiment plats, étaient fabriqués par la maison allemande Heyde et vendus avec de nombreux bateaux Märklin. Les bases de plomb des marins s'encochent sous de petites pattes fixées sur le pont; ceci assure leur stabilite.

Das dampfbetriebene Schlachtschiff „Chicago" wurde zuerst im Katalog von 1904 angeboten. Die Matrosen stammen von der deutschen Firma Heyde und wurden mit vielen Märklin-Schiffen verkauft. Die bleiernen Standplatten der Matrosen konnten zur Erzielung ihrer Standfestigkeit beim Stapellauf unter kleine, auf dem Deck angebrachte Zungen gesteckt werden.

The large torpedo boat 'V 187' is driven by steam. It was illustrated for the first time in the catalogue issued by Märklin in 1904 and was made until the beginning of the First World War; it was re-introduced some time after the war. The steam whistle with wooden control handle is shown on the right between the two funnels.

Le grand torpilleur 'V 187' fonctionne à la vapeur. Il fut illustré pour la première fois dans le catalogue Märklin de 1904 et fabriqué jusqu'au début de la Première Guerre Mondiale; il devait réapparaitre sur le marché quelque temps après la guerre. A droite, entre les deux cheminées, le sifflet à vapeur avec poignée de contrôle en bois.

Das große dampfbetriebene Torpedoboot „V 187" wurde zuerst im Märklin-Katalog von 1904 angeboten und bis Ausbruch des I. Weltkriegs verkauft. Nach dem Krieg kam es erneut auf den Markt. Rechts zwischen den beiden Schornsteinen ist der Holzgriff der Dampfpfeife sichtbar.

This splendid clockwork battleship, made about 1900, measures an impressive 105cm(41⅜in). It is an early example of the series produced up to 1904. The single bar railing and vertical spikes, by which sailors are positioned on deck, indicate its date; deck tabs were not introduced until 1902. Other early features include the figurehead and the deck and porthole painting.

Ce splendide navire de guerre mécanique, fabriqué vers 1900, mesure plus d'un mètre (105cm). C'est l'un des premiers exemplaires de la série produite jusqu'en 1904. La rambarde à barre unique et les pointes verticales par lesquelles on fixait les soldats sur le pont sont indicatrices de la date de fabrication; en effet les pattes sur le pont ne furent pas introduites avant 1902. La figue de proue est une autre caractéristique.

Dieses großartige Aufzieh-Schlachtschiff mißt beachtliche 105cm. Es ist ein um 1900 hergestelltes Beispiel einer bis 1904 produzierten Reihe. Die aus einer Stange bestehende Reling und die vertikalen Dorne zum Aufstecken der Matrosen auf dem Deck verraten das Baujahr, denn Deckzungen kamen erst 1902 auf. Andere frühe Merkmale sind die Galionsfigur und die Bemalung von Deck und Bullaugen.

INA

'Sperber' is run by clockwork; a steam version was also available. Made between 1902 and 1904, it sports many early features including an ornamental brass figurehead. The upper deck is screwed to the hull – an unusual occurrence. This boat is representative of the transition period between the naïve products of the late 19th century and the more sophisticated toys of 1906 to 1914.

'Sperber' est mécanique; une version à vapeur était aussi disponible. Fabriqué de 1902 à 1904 il comprend de nombreuses intéressantes caractéristiques telle la figure de proue en laiton estampé. Le pont supérieur est vissé à la coque – un trait peu commun. Ce bateau est un bon exemple de la transition qui s'opéra entre la production naïve de la fin du 19ème siècle et celle plus sophistiquée des années 1906-1914.

Die „Sperber" hat ein Federwerk, doch gab es auch eine Dampfausführung. Das von 1902 bis 1904 gebaute Schiff hat viele frühe Merkmale wie eine Galionsfigur aus Messing. Ungewöhnlich ist die Verschraubung des Oberdecks mit dem Rumpf. Das Schiff stammt aus der Übergangszeit zwischen den naiven Erzeugnissen des späten 19. Jahrhunderts und den verfeinerten Spielzeugen von 1906 bis 1914.

ERBER

A steam gunboat of 1908, 'Suffren', was available from 1904 to 1912 and was also offered in clockwork form. Despite its small size of 35cm(13⅞in) and its comparatively simple manufacture (apart from hull and gun turrets), it is a typically handsome Märklin toy, complete with original sailors. The accompanying turning crane, for loading stores, was made between 1904 and 1912.

Cette cannonière de 1908, 'Suffren', fut disponible de 1904 à 1912. Elle existait aussi en version mécanique. Malgré sa petite taille de 35cm et la relative simplicité de sa fabrication (à part coque et tourelles) elle possède le charme et la qualité des plus beaux jouets Märklin. Les marins, Märklin aussi, sont d'origine. La grue tournante fut fabriquée entre 1904 et 1912.

Dieses Dampf-Kanonenboot „Suffren" von 1908 gab es zwischen 1904 und 1912 sowie auch mit Federwerk. Trotz der geringen Größe (35cm) und der relativ einfachen Konstruktion (abgesehen von Rumpf und Geschütztürmen) ist es ein typisches, komplett mit Matrosen verkauftes attraktives Märklin-Spielzeug. Der Ladekran wurde zwischen 1904 und 1912 hergestellt.

SUFFREN

The clockwork transatlantic liner 'Carmania' is a good example of Märklin's ability to supply different national markets with toy boats of relevant identity and equipped with suitable accessories, such as national flags. This boat is the smallest of the three sizes offered in the 1909 catalogue: 72cm (28¼in) long.
The well-known original was part of the White Star Line fleet.

Le paquebot transatlantique mécanique 'Carmania' est un bon exemple du talent de Märklin à fournir les différents marchés nationaux en bateaux ayant leur identité distinctive et des accessoires appropriés, comme le drapeau national. Ce bâtiment est le plus petit des trois tailles offertes dans le catalogue de 1909: 72cm. Le célèbre original faisait partie de la flotte White Star Line.

Das Transatlantik-Passagierschiff „Carmania" mit Federwerk ist ein gutes Beispiel für Märklins Fähigkeit, die nationalen Märkte mit echten Vorbildern nachgebauten Schiffen und entsprechender Ausstattung wie z.B. Nationalflaggen zu beliefern. Dieses 72cm lange Schiff ist das kleinste von dreien aus dem Katalog von 1909. Das gut bekannte Vorbild gehörte zur White Star Line.

The 1909 version of the 'Mauretania' shown here is one of Märklin's medium-sized transatlantic liners: 98cm (38½in) long. This impressive boat was available with a clockwork, electric or steam-driven motor. The clockwork version of this size has a running time of twenty minutes, the steam-driven one hour, and the electric seven hours. At 31,938 tons, the original 1907 liner was, in her day, the largest ship in the world; she was also one of the first ships to be driven by turbine engines. Like the 'Carmania', she belonged to the White Star Line (which eventually amalgamated with Cunard). Both 'Mauretania' and 'Carmania' were sister-ships of the 'Lusitania', torpedoed in 1915.

Die hier gezeigte „Mauretania" von 1909 ist mit 98cm eines der mittleren Passagierschiffe Märklins und war mit Federwerk, Dampfantrieb und Elektromotor erhältlich. Die Uhrwerkausführung dieser Größe lief 20 Minuten, die mit Dampfantrieb eine Stunde und die elektrische Version sieben Stunden. Das 1907 gebaute Vorbild war mit 31.938t damals das größte Schiff der Welt und eines der ersten mit Turbinenantrieb. Es gehörte ebenso wie die „Carmania" zur White Star Line (die sich später mit Cunard zusammenschloß). Die „Mauretania" und die „Carmania" waren Schwesterschiffe der 1915 versenkten „Lusitania".

La version de 1909 du 'Mauretania' illustrée ici, figure parmi les paquebots transatlantiques de taille moyenne de Märklin: 98cm de long. Cet impressionnant navire était disponible avec moteur mécanique, électrique ou à vapeur. La version mécanique fontionnait pendant vingt minutes, celle à vapeur pendant une heure, et celle électrique pendant sept heures. Avec ses 31.938 tonnes, l'original de 1907 était en son temps le plus grand navire au monde; il fut aussi l'un des premiers à fonctionner avec des moteur à turbines. Comme le 'Carmania' il appartenait à White Star Line (qui devait plus tard fusionner avec Cunard). Le 'Mauretania' et le 'Carmania' appartenaient à la même compagnie que le Lusitania, torpillé en 1915.

NOVELTIES
NOUVEAUTÉS
NEUHEITEN

This extremely rare aeronautical toy was based on the early Wright brothers' flying machine; it was first illustrated in the 1909 catalogue. A clockwork motor is housed under the pilot's seat; it drives a central spindle, which in turn drives the rubber tyred wheels, the propeller and an overhead pulley wheel. Thus is the plane able to travel along a suspended wire. The propeller and main wing sections are made of celluloid. A particularly attractive feature is the swan's neck tiller which controls the nose wheel and the ailerons.

Ce très rare jouet aéronautique prit modèle sur la 'machine volante' des frères Wright; il fut illustré pour la première fois dans le catalogue de 1909. Un moteur mécanique est situé sous le siège du pilote; il entraine un arbre central qui à son tour entraine les roues à pneus de caoutchouc, l'hélice et une roue de poulie placée au dessus de l'appareil capable ainsi de se déplacer le long d'un fil de fer suspendu. L'hélice et les sections principales des ailes sont en celluloïd. Un trait particulièrement attrayant est le levier en col de cygne qui contrôle la roue avant et les ailerons.

Dieses äußerst seltene Flugspielzeug folgte dem Vorbild der ersten Flugmaschinen der Brüder Wright und wurde zuerst im Katalog von 1909 gezeigt. Ein Federwerk unter dem Pilotensitz treibt über eine Spindel die Gummiräder, den Propeller und eine Riemenscheibe über den Tragflächen an, so daß das Flugzeug sich an einem Spanndraht hängend vorwärtsbewegt. Propeller und Tragflächen bestehen aus Zelluloid. Besonders attraktiv ist der Schwanenhals-Steuerknüppel zur Bedienung von Bugrad und Querrudern.

Märklin issued this fine clockwork Zeppelin in about 1909. The firm's trademark, placed centrally below the Zeppelin's body, is overtly featured, in contrast to the usual discreetly placed logos on other Märklin toys. The propeller is the same size as that on the Wright brothers' biplane: a good example of Märklin's systematic use of the same part for different toys.

Märklin sortit ce beau zeppelin mécanique aux environs de 1909. La marque de la maison apparait de façon proéminente sous le ventre du Zeppelin, en contraste avec les autres jouets Märklin ou elle figure d'habitude de façon plus discrète. L'hélice est de la même taille que celle du biplan des frères Wright: un bon exemple de l'utilisation systématique de la même pièce pour des jouets différents.

Märklin brachte diesen ansehnlichen Aufzieh-Zeppelin um 1909 heraus. Im Gegensatz zu den an anderen Spielzeugen gewöhnlich unauffällig angebrachten Firmenzeichen ist es hier in der Mitte unter dem Rumpf deutlich sichtbar. Der Propeller hat die gleiche Größe wie am Doppeldecker der Brüder Wright – ein Beispiel früher Rationalisierung bei Märklin.

A series of highly decorative miniature baby carriages was featured in the 1909 Märklin catalogue; approximately twenty different styles were produced during this time – a surprisingly wide choice aimed at a relatively small

Une série de très belles voitures d'enfant figurait au catalogue de 1909; environ vingt types différents furent produits à cette époque – choix étonnamment grand pour un marché relativement petit. Les côtés de ce charmant

Der Märklin-Katalog von 1909 enthielt eine Reihe reich verzierter Miniatur-Puppenwagen. Zu jener Zeit wurden ca. 20 Typen hergestellt – eine erstaunliche Anzahl für den relativ kleinen Markt. Die Seitenwände dieses detailgetreu

market. The sides of this delightfully detailed 1909 pram are embossed to resemble basketwork; the handle is behind the lace-trimmed canopy, as opposed to facing it.

landau de 1909 ont été estampés de façon à ressembler à un panier; la poignée est derrière la capote garnie de dentelles au lieu de devant.

gearbeiteten Wagens von 1909 sind zur Imitation von Flechtwerk geprägt. Der Handgriff befindet sich nicht wie üblich am Fuß-, sondern am Kopfende hinter dem mit Spitzen eingefaßten Faltverdeck.

This miniature baby carriage was illustrated in the catalogue issued by Märklin in 1909. It has a particularly original design incorporating adjustable seats which allow two small dolls to be seated facing one another, or

Cette poussette miniature était illustrée dans le catalogue Märklin de 1909. Elle est d'un style particulièrement original en ce qu'elle incorpore des sièges ajustables qui permettent à deux petites poupées d'être assises face à face

Dieser Miniatur-Sportwagen stammt aus dem Märklin-Katalog von 1909. Die originelle Konstruktion besteht aus zwei verstellbaren Sitzen, in denen zwei kleine Puppen entweder Rücken an Rücken oder einander gegenüber

alternatively back to back. The centrally placed fringed canopy gives protection to both sides of this charming double 'pushchair'.

ou dos à dos. La capote au milieu protège des deux côtés cette charmante poussette.

sitzen. Das in der Mitte befestigte Verdeck mit Fransenrand schützt beide Seiten dieses attraktiven Sportwagens.

This fine landau baby carriage was made in about 1909. Its elegant shape and decorative side panels of flowers and foliage were inspired by the Art Nouveau movement. Complete with a collapsable ruffled satin hood and curved handlebars, this baby carriage is pushed from the front.

Ce joli landau fut fabriqué aux alentours de 1909. Sa forme élégante et ses panneaux latéraux décorés de fleurs et feuillage furent inspirés par l'Art Nouveau.
Ce landau était fourni avec une capote pliante en satin plissé et des poignées recourbées. On le poussait par devant.

Dieser hübsche Kinderwagen im Landauerstil wurde um 1909 hergestellt. Bei der eleganten Form und dem Blumen- und Blattwerkmuster an den Seitenwänden stand die Art Nouveau-Bewegung Pate. Ein Satinfaltverdeck und eine gebogene Griffstange runden den Wagen ab.

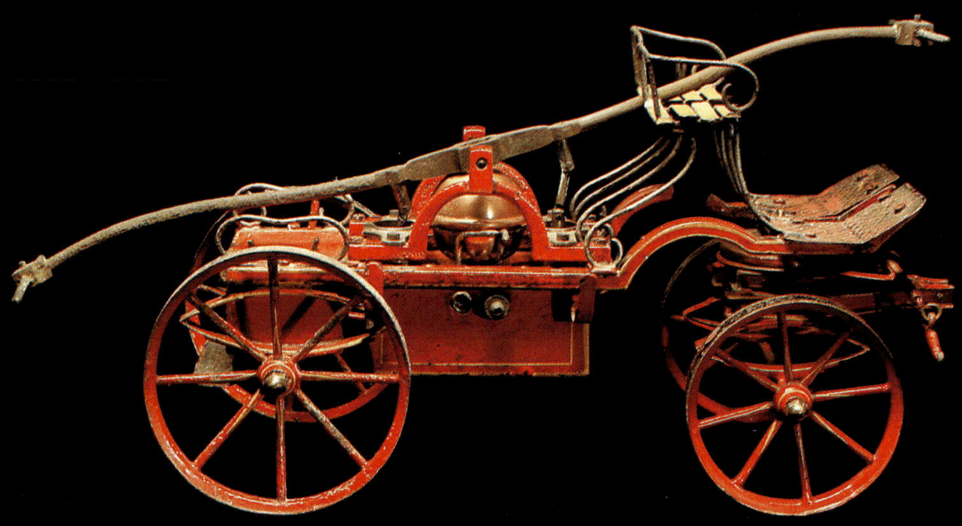

Horse-drawn and hand-operated, this interesting fire pump was marketed from 1902 to 1912. It is a strongly constructed toy with a cast-iron chassis and a realistic working mechanism. In common with many working toys made by Märklin during this period, the main parts of the fire pump are bolted or screwed together. The delicately shaped slatted seats are, however, soldered onto the main chassis (as can be seen in the bottom left-hand detail). The original bell, lamps and hose-reel are missing from this fire pump.

Cette curieuse pompe à incendie, attelée à un cheval et que l'on opérait manuellement fut vendue entre 1902 et 1912. C'est un jouet construit solidement avec un chassis en fonte et un mécanisme qui fonctionnait réellement. En commun avec de nombreux jouets de ce type fabriqué par Märklin pendant cette période, les composantes principales de la pompe sont boulonnées ou vissées ensemble. Les délicats sièges à lames sont néanmoins soudés au chassis (ainsi qu'il est visible sur le détail à gauche).
Sont manquants sur cet exemplaire la cloche, les lanternes et le rouleau de tuyau d'incendie d'origine.

Dieser interessante, von 1902 bis 1912 verkaufte Spritzenwagen wurde von Pferden gezogen und von Hand bedient. Auf dem stabilen gußeisernen Fahrgestell befindet sich ein detailgetreuer arbeitsfähiger Mechanismus. Wie viele von Märklin zu jener Zeit gebaute Arbeitsmodelle sind die Hauptteile miteinander verschraubt, und nur die fein gearbeiteten Lattensitze sind (wie der linke Ausschnitt zeigt) mit dem Fahrgestell verlötet. An diesem Modell fehlen die Feuerglocke, die Lampen und die Schlauchrolle.

This horse-drawn coupé dates from 1909. Earlier models, produced from 1900, were equipped with a naïve tinplate horse as opposed to the authentically harnessed hide-covered type shown here, which was available towards the end of the decade. This transition is typical of Märklin's attempts to achieve greater realism, a trend that became more strongly marked after the war.

Cette calèche à cheval date de 1909. Les modèles précédents, produits depuis 1900, étaient équipés avec un simple cheval en fer-blanc bien différent de celui présenté ici, recouvert de cuir et harnaché de façon authentique, qui fut disponible vers la fin de la décennie. Cette transition est typique des efforts de Märklin à créer un réalisme plus grand, tendance qui deviendra plus fortement marquée après la guerre.

Diese Pferdekutsche von 1909 besaß im Gegensatz zu den nach 1900 gefertigten Modellen mit ihren einfachen Blechpferden ein originalgetreu angeschirrtes und mit Fell überzogenes Pferd. Dieser gegen Ende des ersten Jahrzehnts erhältliche Typ demonstriert Märklins Streben nach realistischeren Nachbildungen, das sich nach Kriegsende noch verstärkte.

Märklin marketed this beautiful fairground carousel in about 1909. The toy, which has a simple musical movement, can either be cranked by hand or used as a steam engine accessory.

Märklin mit en vente ce joli manège aux alentours de 1909. Le jouet, qui possède un mouvement mécanique simple, peut être mis en route à la manivelle ou utilisé comme accessoire de machine à vapeur.

Märklin verkaufte dieses entzückende Jahrmarktskarussell um 1909. Das mit einer einfachen Spieluhr versehene Karussell wurde entweder mit einer Handkurbel oder von einer Dampfmaschine gedreht.

The minutely detailed hand-painting of the carousel and its original plaster figures can be seen here in close-up. The amusing seats, embossed in shell shapes, the central hexagonal column, decorated with mirrors, painted panels and medallions, and the intricately shaped metalwork are some of the features which make this toy especially appealing.

La peinture minutieuse et détaillée du manège et ses personnages d'origine en plâtre peuvent être vus ici en gros-plan. Parmi les traits qui rendent ce jouet particulièrement attrayant on remarque les sièges pittoresques, estampés de motifs de coquilles, la colonne hexagonale centrale décorée de miroirs, médaillons et paneaux peints, et la serrurerie finement ouvragée.

Die sorgfältig ausgeführte Handbemalung des Karussells und der Gipsfiguren ist in der Vergrößerung deutlich zu erkennen. Dieses Spielzeug gewinnt speziell durch die muschelförmigen Sitze, die sechseckige Mittelsäule mit ihren Spiegeln, Verzierungen und Porträts und die fein ausgeführten Drahtarbeiten.

One of the very few novelty toys made by Märklin, this rare 1909 clockwork circus floor-train, known as 'Fidelitas', is equipped with a large cam which enables it to move forwards in an irregular pattern. Its wheels are identical to

Parmi les quelques nouveautés fabriquées par Märklin on trouve ce rare train de cirque mécanique de 1909, connu sous le nom de 'Fidelitas', équipé d'une grande came qui lui permet d'avancer de façon irrégulière. Ses roues sont

Dieser seltene als „Fidelitas" bekannte Zirkuszug von 1909 mit Federwerk war eine der wenigen Neuheiten Märklins. Eine große Nocke bewirkte das Fahren in ständig wechselnden Richtungen. Die Räder sind die gleichen

those found on some of Märklin's contemporary baby carriages. The quality of the painting is remarkable for a factory produced toy.

identiques à celles des voitures de poupées de la même époque. La qualité de la peinture est remarquable pour un jouet fabriqué en usine.

wie bei einigen Puppenwagen Märklins aus der gleichen Zeit. Die Güte der Lackierung ist für ein Serienerzeugnis eine erstaunliche Leistung.

Fidelitas

In contrast to other manufacturers, whose product boxes often featured strong graphics, Märklin usually used unadorned wooden or cardboard boxes. The packaging of these 1912 spinning tops is an exception.

Au contraire des autres fabricants dont les coffrets étaient souvent très décorés, ceux de Märklin n'étaient souvent que de simples boites en boisblanc ou en carton. Cette boîte de toupies de 1912 est une exception.

Im Gegensatz zu anderen Herstellern, deren Verpackungen meist stark bedruckt und verziert waren, bevorzugte Märklin schlichte Holz- oder Pappschachteln. Diese Kreiselschachtel von 1912 ist eine Ausnahme.

The evocative full colour illustration on the boxlid of a 1908 Midland Railway gauge 0 steam locomotive is another unusual example of the use of decorative graphics on Märklin packaging.

Cette illustration évocatrice en couleur sur le couvercle d'un coffret d'une locomotive Midland Railway à vapeur de 1908 en écartement 0 est un autre exemple inhabituel de l'emploi de motifs décoratifs.

Diese verführerische farbige Schachtel für eine Midland Railway-Dampflok der Spur 0 von 1908 ist ein ungewöhnliches Beispiel für die dekorative Gebrauchsgraphik auf Märklins Verpackungen.

One of the more bizarre toys offered by Märklin between 1904 and 1910 was this windmill. It has a strange and original action; when one of its sails is hit by an arrow, the building 'explodes'. The collapse that occurs as a result of the arrow hit is shown on the right.

Ce moulin était l'un des jouets les plus étranges offerts par Märklin entre 1904 et 1910. Il a un fonctionnement très original: lorsqu'une aile est touchée par une flèche, le moulin 'explose'. Le résultat de 'l'explosion' est illustré à droite.

Diese seltsame Windmühle wurde von Märklin zwischen 1904 und 1910 angeboten. Ungewöhnlich ist auch ihre Arbeitsweise: Wenn ein Flügel von einem Pfeil getroffen wird, „explodiert" sie. Das rechte Bild zeigt die zerfallene Mühle nach einem Treffer.

The manufacture of miniature cooking stoves and accessories was Märklin's first line of business. Both these 1909 cookers are fired by methylated spirit burners. The cast-iron doors of the stove on the left are nickel-plated, as they would have been on the original. Some beautifully modelled accessories stand on the stoves' burners.

La fabrication de fourneaux et accessoires de cuisine fut la première spécialité de Märklin. Ces deux cuisinières fonctionnent à l'alcool. Les portes en fonte du réchaud de gauche sont nickelées ainsi qu'elles l'auraient été sur l'original. Quelques beaux accessoires sont placés sur les brûleurs.

Die ersten Erzeugnisse Märklins waren Puppenherde und Zubehör. Diese beiden Herde von 1909 werden mit einer Spiritusflamme betrieben. Die Gußeisentüren des linken Herdes sind wie die echten Vorbilder vernickelt. Auf der Ofenplatte stehen einige entzückend nachgebildete Töpfe und Kessel.